Advances in Pharmaceutical Sciences

Volume 7

Advances in Pharmaceutical Sciences

Volume 7

edited by

David Ganderton
King's College, London

Trevor Jones
The Association of the British Pharmaceutical Industry, London

and

James McGinity
College of Pharmacy, The University of Texas at Austin, USA

ACADEMIC PRESS

Harcourt Brace and Company, Publishers

London · San Diego · New York · Boston · Sydney
Tokyo · Toronto

ACADEMIC PRESS LIMITED
24–28 Oval Road
London NW1 7DX

United States Edition published by
ACADEMIC PRESS INC.
San Diego, CA 92101

A catalogue record for this book is available from the British Library
ISBN 0-12-032307-9

Transferred to digital printing 2006

Filmset by Colset Pte Ltd
Printed and bound by CPI Antony Rowe, Eastbourne

CONTRIBUTORS

J. G. Nairn *Faculty of Pharmacy, University of Toronto, Toronto, Ontario, Canada*

H. G. Kristensen *Department of Pharmaceutics, The Royal Danish School of Pharmacy, Copenhagen, Denmark*

R. J. Roberts *ZENECA Pharmaceuticals, Macclesfield, Cheshire, UK*

R. C. Rowe *ZENECA Pharmaceuticals, Macclesfield, Cheshire, UK*

A. Urtti *Department of Pharmaceutical Technology, University of Kuopio, P.O. Box 1627, 70211 Kuopio, Finland*

CONTENTS

PREFACE

Like its predecessors, Volume 7 of *Advances in Pharmaceutical Sciences* presents critical evaluations of current research and development in selected fields of pharmaceutical technology. It reflects, in particular, two activities vital to the industry. The first is the provision of a better understanding of well-established processes. This provides both a rationale for further exploitation and a secure basis for the management of quality. The contributions of Rowe, Roberts and Kristensen, all world authorities in the field of powder technology, exemplify this principle with major contributions which displace much of the empiricism associated with the characterization of powders and agglomeration.

The second major contribution of pharmaceutical technology lies in refined drug delivery. This is founded on a mastery of physico-chemical principles which are then related to clinical practice through pharmacokinetics. The first part of this sequence is illustrated in Nairn's comprehensive review, which assesses the considerable contribution of coacervation techniques, and the second by Urtti's account of the pharmacokinetics of ocular drug delivery. These accounts provide a strong base for rational product development and future invention.

D. Ganderton
Kings College, London

T. M. Jones
The Association of the British Pharmaceutical Industry

J. W. McGinity
The University of Texas at Austin

1

THE MECHANICAL PROPERTIES OF POWDERS

R. C. Rowe and R. J. Roberts

ZENECA Pharmaceuticals, Macclesfield, Cheshire, UK

INTRODUCTION

The vast majority of drugs, when isolated, exist as crystalline or amorphous solids. Subsequently, they may be either milled (comminuted) and/or admixed with other inactive solids (excipients) and finally filled into capsules or compacted to form tablets. The processes of comminution and compaction involve subjecting the materials to stresses that cause them to undergo deformation. The reaction of the material to the deformation stress, σ_d, is dependent on both the mode of deformation and the mechanical properties of the material.

(a) For elastic deformation

$$\sigma_d = \epsilon E \tag{1}$$

where E is the Young's modulus of elasticity of the material and ϵ is the deformation strain.

(b) For plastic deformation

$$\sigma_d = \sigma_y \tag{2}$$

where σ_y is the yield stress of the material.

(c) For brittle fracture

$$\sigma_d = \frac{AK_{IC}}{\sqrt{d}} \tag{3}$$

where K_{IC} is the critical stress intensity factor of the material (an indication of the stress required to produce catastrophic crack propagation), d is the particle size (diameter) and A is a constant depending on the geometry

1

ADVANCES IN PHARMACEUTICAL SCIENCES
ISBN 0–12–032307–9

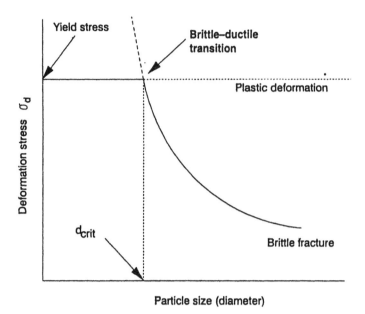

Fig. 1 Schematic diagram showing the effect of particle size on the deformation stress for materials that undergo brittle fracture and/or plastic deformation.

and stress application. For compression of rectangular samples with large cracks $A = \sqrt{32/3}$ or 3.27 (Kendall, 1978) but for other geometries A varies between 50 and 1 (Puttick, 1980).

It is evident from the equations and Fig. 1 that large particles will tend to crack because the stress required for brittle fracture will be less than that needed for plastic flow and vice versa for very small particles. The transition from brittle to ductile behaviour will occur at a critical size, d_{crit}, where the two stresses will be equal, i.e.

$$d_{crit} = \left(\frac{AK_{IC}}{\sigma_y}\right)^2 \tag{4}$$

Application of equation 4 will be discussed later in the chapter.

It is evident from the discussion above that in order to be able to predict the comminution and/or compaction behaviour of a material it is essential that methods be derived to measure for powdered materials:

(a) The Young's modulus of elasticity (E)

(b) The yield stress (σ_y) – this is directly related to the indentation hardness, H, since for a plastic material:

$$\sigma_y = \frac{H}{3} \tag{5}$$

(c) The critical stress intensity factor (K_{IC}) – this is directly related to the fracture toughness, R, since for plane stress:

$$K_{IC} = (ER)^{1/2} \tag{6}$$

In this chapter various methods that have been specifically applied to pharmaceutical materials are reviewed and critically examined.

YOUNG'S MODULUS OF ELASTICITY

If an isotropic body is subjected to a simple tensile stress in a specific direction it will elongate in that direction while contracting in the two lateral directions; its relative elongation is directly proportional to the stress. The ratio of the stress to the relative elongation (strain) is termed Young's modulus of elasticity. This is a fundamental property of the material which is directly related to its interatomic or intermolecular binding energy for inorganic and organic solids, respectively, and is a measure of its stiffness.

The Young's modulus of elasticity of a material can be determined by many techniques, several of which have been used in the study of pharmaceutical materials, namely, flexure testing using both four- and three-point beam bending, indentation testing on both crystals and compacts, compression testing, the split Hopkinson bar configuration and measurements from other moduli.

Flexure testing (beam bending)

In flexure testing, a rectangular beam of small thickness and width in comparison with its length is subjected to transverse loads and its central deflection caused by bending is measured. The beam may be supported and loaded in one of two ways (Fig. 2). If the beam is supported at two points and is loaded at two points it undergoes what is known as four-point bending while if it is supported at two points but is loaded at one point it undergoes what is known as three-point bending. In both systems the beam must be supported symmetrically with an overhang at both ends and the loading points must be symmetrical about the mid-point of the specimen. Equations for the calculation of Young's modulus from the applied load F and the deflection of the mid-point of the beam ξ can easily be derived, e.g. for four-point bending (Church, 1984):

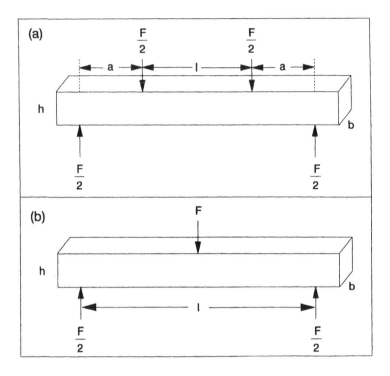

Fig. 2 Geometries for (a) four-point and (b) three-point beam bending. F, applied load; h, beam thickness; b, beam width; l and a, distances between loading points as shown.

$$E = \frac{F}{\xi} \frac{6a}{h^3b} \left(\frac{l^2}{8} + \frac{al}{2} + \frac{a^2}{3} \right) \tag{7}$$

for three-point beam bending (Roberts *et al.*, 1991):

$$E = \frac{Fl^3}{4\xi h^3 b} \tag{8}$$

where h and b are the height (thickness) and width (breadth) respectively and l and a are as given in Fig. 2. In both cases the value for Young's modulus for the specimen under test can either be calculated from a single point determination or more commonly from the slope of total load (F) versus central deflection (ξ).

The four-point beam bending test was first used for pharmaceutical materials by Church (1984) and has since been adapted by Mashadi and

Newton (1987a,b), Bassam *et al.* (1988, 1990, 1991), Roberts *et al.* (1989b) and Roberts (1991). Generally beams of varying height have been used – 100 mm long by 10 mm wide or 60 mm long by 7 mm wide – and are prepared using specially designed punches and dies (Plate 1). Beams are generally prepared at varying compression pressures to achieve specimens of varying porosity. Although the height of the beam (at constant porosity) is an experimental variable, Bassam *et al.* (1990) has shown that it does not have a significant effect on the measured modulus.

The testing rig used by these workers consists essentially of three parts (Plate 2): an upper frame (A) with a platform on which the beam is supported on the two outer contact points and which also holds a displacement transducer kept in contact with the lower surface of the beam by means of an elastic band, a block (B) containing the inner contact points resting on the upper surface of the beam, and a lower frame (C) located in the centre of block (B) by means of a ball bearing. In all cases the contact points are cylindrical and mounted in ball-bearing races to allow free movement. The two frames are attached to the moving platens of a tensile-testing machine which allows measurements to be made at varying loading rates. Generally, however, low rates are used (≈ 1 mm min^{-1}) but Bassam *et al.* (1990) have shown independence of loading rate up to rates of 15 mm min^{-1}.

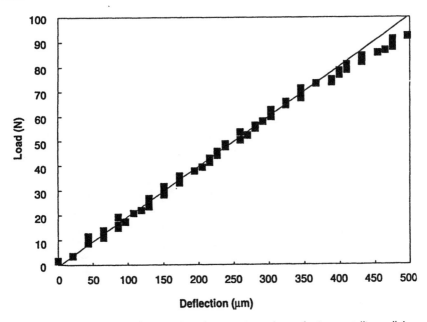

Fig. 3 A typical load–deflection curve for a beam 100 mm long of microcrystalline cellulose.

A typical load–deflection curve for a beam 100 mm long of microcrystal-line cellulose compacted to a porosity of 14.6% is shown in Fig. 3. Gen-erally, each beam is tested five times and for the specimen shown in Fig. 3 calculations of the Young's modulus of elasticity gave a value of 3.53 ± 0.07 GPa, i.e. a coefficient of variation of 2%; such reproducibility is not uncommon with this method of measurement.

A disadvantage of the four-point beam test method is that it invariably requires large specimens and hence large quantities of materials (15–20 g). In addition, high-tonnage presses are needed to prepare the specimens thus exacerbating problems with cracking and lamination on ejection from the die. It was to overcome these difficulties that Roberts *et al.* (1989b) deve-loped a three-point beam testing method that uses beams prepared from 200 mg of material. The beams in this case are 20 mm long by 7 mm wide and are stressed by applying a static load of 0.3 N with an additional

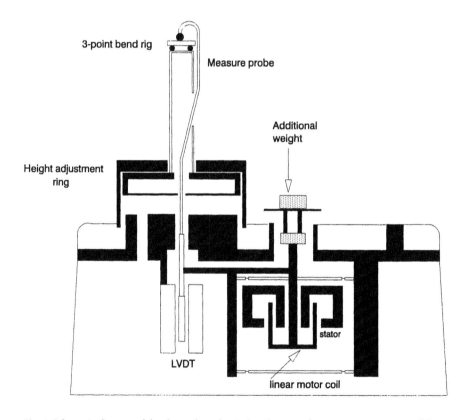

Fig. 4 Schematic diagram of the thermal mechanical analyser used to measure Young's modulus.

dynamic load of $\pm 0.25\,N$ (at a frequency of $0.17\,Hz$) using a thermal mechanical analyser (Mettler Instruments TMA40). The schematic diagram (Fig. 4) shows the position of the linear differential displacement transducer attached to the measuring probe and the modification made to enable the application of extra loading while Plate 3 shows the three-point bending rig with the measuring probe. In operation a calibration run is first performed to eliminate distortions in the sensor and other parts of the displacement measuring system and then 20 measurements of specimen displacement are undertaken to an accuracy of $\pm 0.005\,\mu m$. The Young's modulus of elasticity of the specimen is then calculated from the mean displacement corrected for distortions using equation 8, where in this case F is the applied dynamic load. Extensive testing by Roberts *et al.* (1989b) and Roberts (1991) has shown equivalence between this test and the conventional four-point beam test despite the fact that it is generally accepted that in the latter there is a more uniform stress pattern over the central section of the beam and little contribution from shear stresses.

A problem associated with the analysis of data for specimens prepared from particulate solids is in the separation of the material property from that of the specimen property which by definition includes a contribution by the porosity of the specimen. All workers have found that for all materials there is a decrease in Young's modulus with increasing porosity (Fig. 5). Numerous equations have been published which describe this relationship (Dean and Lafez, 1983). Certain equations are based on theoretical considerations (Wang, 1984; Kendall *et al.*, 1987) while others are empirical curve-fitting functions (Spriggs, 1961; Spinner *et al.*, 1963). Recently, Bassam *et al.* (1990) have reported a comparison of all the equations currently in use for data generated from four-point beam bending on 15 pharmaceutical powders ranging from celluloses and sugars to inorganic materials such as calcium carbonate and have concluded that the best overall relationship is the modified two-order polynomial (Spinner *et al.*, 1963)

$$E = E_o(1 - f_1 P + f_2 P^2) \qquad (9)$$

where E_o is the Young's modulus at zero porosity and E is the measured modulus of the beams compacted at porosity, P; f_1 and f_2 are constants. However such a conclusion may not be universal as during extensive studies on a wider range of materials including drugs, Roberts (1991) concluded that an exponential relationship (Spriggs, 1961) is the preferred option for data generated using the three-point beam testing method, i.e.

$$E = E_o \exp^{-bP} \qquad (10)$$

where b is a constant. It is interesting to note that, on average, the extrapolated values of E_o (Young's modulus at zero porosity) calculated using

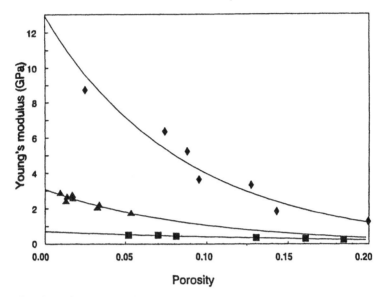

Fig. 5 The effect of porosity on the measured Young's modulus of ■, PTFE; ▲, testosterone propionate; ♦, theophylline (anhydrous). Data generated using three-point beams (taken from Roberts *et al.*, 1991).

both equations are only marginally different (Bassam *et al.*, 1988, 1990).

The analysis of Young's modulus at zero porosity thus provides a means of quantifying and categorizing the elastic properties of powdered materials. Table 1 shows literature data for a variety of pharmaceutical excipients and drugs determined using beam bending methods. It can be seen that the values vary over two orders of magnitude ranging from hard rigid materials with very high moduli (e.g., the inorganics) to soft elastic materials with low moduli (e.g., the polymeric materials). As a result a rank order of increasing rigidity of tabletting excipients can be listed: starch < microcrystalline celluloses < sugars < inorganic fillers with variations in the groups dependent on chemical structure as well as the preparation and pretreatment routes (including particle size).

The effects of particle size are distinguishable within the celluloses and α-lactose monohydrate. In the former there is a small increase with decreasing particle size while in the latter the increase is much greater. Whereas in the former the effect is probably due to an increase in contact area, in the latter the effect is due to specimen defects in that the specimens used by Bassam *et al.* (1990) contained microscopic flaws and cracks. Recent work by Roberts (1991) has shown that it is necessary to eliminate all specimens with cracks otherwise the extrapolated modulus values to zero porosity are inconsistent.

For the lactose samples the rank order of increasing rigidity is spray dried < β-anhydrous < α-monohydrate, consistent with the findings of workers describing the compaction properties of the materials using instrumented tabletting machines (Fell and Newton, 1971; Vromans *et al.* 1986). The variations in the cellulose samples can be attributed to subtle differences in the manufacturing preparative technique. However, it is known that this factor can also affect the equilibrium moisture content of the samples with Unimac samples attaining a lower equilibrium moisture content than Avicel samples. As it is known that increasing moisture content can lead to a decrease in Young's modulus for microcrystalline cellulose (Bassam *et al.*, 1990) the differences in modulus between sources listed in Table 1 would be expected to increase if all materials were compared at equivalent moisture contents.

All the drugs tested exhibited low moduli equivalent to the polymeric materials. This is not totally unexpected as it is known that many organic solids including drugs form glasses which exhibit anomalous endotherms that resemble glass transitions and can therefore be regarded as possessing a certain amount of mobility.

Recently, Roberts *et al.* (1991) have investigated the relationship between the Young's modulus of a variety of drugs and excipients using three-point beam bending and their molecular structure based on intermolecular interactions using the concept of cohesive energy density (CED). They found a direct relationship of the form

$$E_o = 0.01699\text{CED} - 2.7465 \qquad (11)$$

where CED is expressed in units of MPa and Young's modulus in units of GPa. This equation compares favourably with that derived from the Tobolsky (1962) equation relating the bulk modulus, K, of a face-centred cubic lattice at $0°$ K to the cohesive energy density:

$$K = 0.0141\text{CED} \qquad (12)$$

Since K is related to E by the equation:

$$K = \frac{E}{3(1 - 2v)} \qquad (13)$$

where v is the Poisson's ratio (taken at 0.3 for the majority of pharmaceutical materials) then combining equations 12 and 13 gives:

$$E_o = 0.01692\text{CED} \qquad (14)$$

Table 1 Young's modulus at zero porosity measured by flexure testing

Material	Method	Young's modulus (GPa)	Particle size (μm)	Reference
Celluloses				
Avicel PH101	4PB (100 × 10) EXP	10.3	50	Mashadi and Newton (1987b)
Avicel PH101	4PB (100 × 10) EXP	9.7	50	Roberts and Rowe (1987c)
Avicel PH101	4PB (100 × 10) POLY	9.2	50	Bassam *et al.* (1990)
Avicel PH101	4PB (100 × 10) EXP	9.0	50	Bassam *et al.* (1988)
Avicel PH101	3PB (20 × 7) EXP	7.8	50	Roberts *et al.* (1989b)
Avicel PH101	4PB (100 × 10) EXP	7.6	50	Roberts *et al.* (1989b)
Avicel PH101	4PB (60 × 7) EXP	7.4	50	Roberts *et al.* (1989b)
Avicel PH102	4PB (100 × 10) POLY	8.7	90	Bassam *et al.* (1990)
Avicel PH102	4PB (100 × 10) EXP	8.2	90	Bassam *et al.* (1988)
Avicel PH105	4PB (100 × 10) EXP	10.1	20	Bassam *et al.* (1988)
Avicel PH105	4PB (100 × 10) POLY	9.4	20	Bassam *et al.* (1990)
Emcocel	4PB (100 × 10) EXP	9.0	56	Bassam *et al.* (1988)
Emcocel	4PB (100 × 10) POLY	7.1	56	Bassam *et al.* (1990)
Emcocel (90M)	4PB (100 × 10) EXP	9.4	90	Bassam *et al.* (1988)
Emcocel (90M)	4PB (100 × 10) POLY	8.9	90	Bassam *et al.* (1990)
Unimac (MG100)	4PB (100 × 10) EXP	8.8	38	Bassam *et al.* (1988)
Unimac (MG100)	4PB (100 × 10) POLY	8.0	38	Bassam *et al.* (1990)
Unimac (MG200)	4PB (100 × 10) EXP	8.0	103	Bassam *et al.* (1988)
Unimac (MG200)	4PB (100 × 10) POLY	7.3	103	Bassam *et al.* (1990)
Elcema (P100)	4PB (100 × 10) EXP	8.6	–	Roberts and Rowe (1987c)
Sugars				
Sorbitol instant	4PB (100 × 10) EXP	45.0	–	Mashadi and Newton (1987a)
α-Lactose monohydrate	3PB (20 × 7) EXP	24.1	20	Roberts *et al.* (1991)
α-Lactose monohydrate	4PB (100 × 10) POLY	3.2	63	Bassam *et al.* (1990)
Lactose β anhydrous	4PB (100 × 10) POLY	17.9	149	Bassam *et al.* (1990)
Lactose β anhydrous	4PB (100 × 10) POLY	18.5	149	Bassam *et al.*(1991)

Lactose (spray dried)	4PB (100 × 10) EXP	13.5	–	Roberts and Rowe (1987c)
Lactose (spray dried)	4PB (100 × 10) POLY	11.4	125	Bassam et al. (1990)
Dipac sugar	4PB (100 × 10) POLY	13.4	258	Bassam et al. (1990)
Mannitol	4PB (100 × 10) POLY	12.2	88	Bassam et al. (1990)
Polysaccharides				
Starch 1500	4PB (100 × 10) EXP	6.1	–	Roberts and Rowe (1987c)
Maize starch	4PB (100 × 10) POLY	3.7	16	Bassam et al. (1990)
Inorganics				
Emcompress	4PB (100 × 10) EXP	181.5	–	Roberts and Rowe (1987c)
Calcium carbonate	4PB (100 × 10) POLY	88.3	8	Bassam et al. (1990)
Calcium phosphate	4PB (100 × 10) POLY	47.8	10	Bassam et al. (1990)
Polymers				
PVC	4PB (100 × 10) POLY	4.4	–	Bassam et al. (1991)
PVC	3PB (20 × 7) EXP	4.1	–	Roberts et al. (1991)
Stearic acid	3PB (20 × 7) EXP	3.8	62	Roberts et al. (1991)
PTFE	4PB (100 × 10) EXP	0.81	–	Roberts et al. (1989b)
PTFE	4PB (100 × 10) POLY	0.71	–	Bassam et al. (1991)
PTFE	3PB (20 × 7) EXP	0.71	–	Roberts et al. (1989b)
Drugs				
Theophylline (anhydrous)	3PB (20 × 7) EXP	12.9	31	Roberts et al. (1991)
Paracetamol DC	3PB (20 × 7) EXP	11.7	120	Roberts (1991)
Caffeine (anhydrous)	3PB (20 × 7) EXP	8.7	38	Roberts et al. (1991)
Sulphadiazine	3PB (20 × 7) EXP	7.7	9	Roberts et al. (1991)
Aspirin	3PB (20 × 7) EXP	7.5	32	Roberts et al. (1991)
Ibuprofen	3PB (20 × 7) EXP	5.0	47	Roberts et al. (1991)
Phenylbutazone	3PB (20 × 7) EXP	3.3	50	Roberts et al. (1991)
Testosterone propionate	3PB (20 × 7) EXP	3.2	85	Roberts et al. (1991)

PTFE, polytetrafluoroethylene; PVC, polyvinyl chloride; 4PB = four-point beam; 3PB = three-point beam; EXP = equation 10; POLY = equation 9; the beam dimensions (length × width in mm) are given in parentheses.

It can be seen that the moduli predicted from this equation are somewhat higher than those measured (Fig. 6). This is thought to be due to the fact that all the measurements were carried out at 298° K and hence the measured modulus would be lower. The significance of this finding is in the recognition of the validity of both the test method and the data manipulation (i.e., the extrapolation to zero porosity).

Indentation testing

In indentation testing a hard indenter made of either diamond, sapphire or steel and machined to a specific geometry – either a square based pyramid (Vickers indenter) or a spherical ball (Brinell indenter) – is pressed under load into the surface of a material either in the form of a crystal or compacted specimen. Although well recognized as a method for measuring hardness (see later) it may also be used to measure Young's modulus except that in this case it is the recovered depth after removal of the load that is important (Fig. 7). The test may be either static (Ridgway *et al.*, 1970; Duncan-Hewitt and Weatherly, 1989a,b) or dynamic involving a pendulum (Hiestand *et al.*, 1971).

For crystals a Vickers indenter is generally used and in this respect Duncan-Hewitt and Weatherly (1989a,b) have used a Leitz–Wetzler Mini-

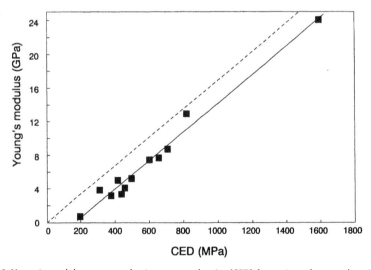

Fig. 6 Young's modulus versus cohesive energy density (CED) for various drugs and excipients; - - - - theoretical line, equation 14 (taken from Roberts *et al.*, 1991).

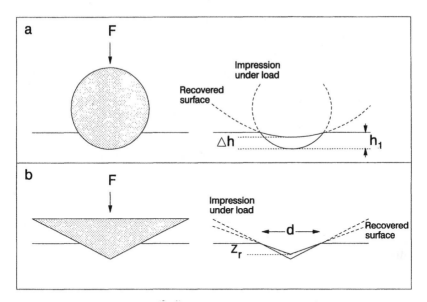

Fig. 7 Loading geometry and recovery of indent for (a) Brinell and (b) Vickers hardness testers used to measure Young's modulus. F, applied load. For definitions of other terms, see text.

load hardness tester, applying a load of 147 mN. Single crystals were pre-selected and were mounted in plasticine with the surface normal to the indentation direction. Load was applied over 15 s and was maintained for 10 s. In the case of sucrose crystals, hardness anisotropy of the various faces was demonstrated (see later).

The equation used by Duncan-Hewitt and Weatherly (1989b) involves measuring the recovered depth of indentation and the length of the Vickers diagonal and substituting in the equation of Breval and Macmillan (1985):

$$\left(\frac{2Z_r}{d}\right)^2 = 0.08168\left[1 - 8.7\left(\frac{H_v}{E}\right) + 14.03\left(\frac{H_v}{E}\right)^2\right] \qquad (15)$$

where H_v is the indentation hardness (see later for equation), Z_r is the recovered depth of the indentation and d is the length of the Vickers diagonal (Fig. 7).

Data on several materials (Table 2) show values of the same order as those determined by flexure testing. The value for sodium chloride compares favourably with that reported in the literature (37 GPa; Simmons and Wang, 1971).

For compacts Ridgway *et al.* (1970) used a pneumatic microindentation apparatus to measure both the initial depth of penetration (h_1) and

Table 2 Young's modulus measured by indentation on crystals

Material	Young's modulus (GPa)	Method	Reference
Sodium chloride	43.0	Vickers	Duncan-Hewitt and Weatherly (1989b)
Sucrose	32.3	Vickers	Duncan-Hewitt and Weatherly (1989b)
Paracetamol	8.4	Vickers	Duncan-Hewitt and Weatherly (1989b)
Adipic acid	4.1	Vickers	Duncan-Hewitt and Weatherly (1989b)
Aspirin	0.32	Brinell	Ridgway *et al.* (1970)

recovery (Δh) after application of small loads (4 g) using a 1.5 mm diameter sapphire ball indenter and substituting in a modified Hertz equation (Aulton, 1977):

$$E(\text{GPa}) = 1.034 \frac{F}{\Delta h \sqrt{h_1}} \qquad (16)$$

where both h_1 and Δh are measured in micrometres and F, the indentation load, is measured in grams. The values quoted for aspirin are very small, two orders of magnitude below the figure given in Table 1 and hence must be viewed with some suspicion.

In the dynamic method used by Hiestand *et al.* (1971) a 1 m long pendulum with a steel sphere of 25.4 mm diameter strikes the face of a compact of cross-sectional area of 14.52 cm^2. The modulus of elasticity can be calculated from a knowledge of the indentation hardness, H_b, and the strain index, ξ_i, a measure of relative strain during elastic recovery that follows plastic deformation using the equation:

$$E = \frac{H_b(1 - v^2)}{\xi_i} \qquad (17)$$

where v is Poisson's ratio. The strain index, ξ_i, is calculated using values from the indentation experiment:

$$\xi_i = \frac{5a}{6\pi r} \left(\frac{h_i}{h_r} - \frac{3}{8} \right)^{-1} \qquad (18)$$

where a is the chordal radius of the indent, r is the radius of the spherical indenter, h_i and h_r are the initial and rebound heights of the indenter respectively.

Assuming a Poisson's ratio of 0.3 it is possible, using the calculated values for hardness and strain index (Hiestand, 1985), to determine the Young's modulus for spray-dried lactose and microcrystalline cellulose (Avicel PH102): 25.5 and 6.2 GPa, respectively. These values compare favourably with those obtained from flexure testing (Table 1).

Compression testing

In compression testing a compressive stress is applied to either a crystal or compacted specimen and the corresponding strain measured, the ratio of the stress to strain being a measure of the compressive Young's modulus of elasticity.

The loading system depends on the size of specimen, ranging from a micro tensile testing instrument providing a load up to 5 N and a minimum displacement of 5 nm, as used for single crystals by Ridgway *et al.* (1969a), to an Instron physical testing instrument providing a load up to 50 kN, as used for flat-faced cylindrical compacts (8 mm diameter) by Kerridge and Newton (1986). The Young's modulus in compression can be determined using the following equation:

$$E = \frac{L}{(X - C)A} \tag{19}$$

where L is the length of the specimen, X is the slope of strain versus stress, C is the machine constant (strain versus stress for loading of the machine without a specimen) and A is the cross-sectional area of the specimen.

While for crystals single measurements are all that is necessary, for compacts measurements on specimens prepared at different porosities are required. In the latter test the Young's modulus at zero porosity can be determined using equation 10.

Values for the compressive modulus of a variety of materials in Table 3 are lower than those measured by other techniques although the trend of decreasing modulus with increasing particle size is the same as that reported from flexure testing (Bassam *et al.*, 1990). Furthermore, for the tests involving crystals Ridgway *et al.* (1969a) indicated that cracks were important and they attributed this factor to variation in the results. This factor may also account for the low values of modulus when compared with those from flexure testing. In view of this and the fact that the compressive modulus should always be greater than the tensile modulus, the results from these tests must be viewed with some scepticism.

Table 3 Young's modulus measured by compaction testing

Material	Method	Particle size (μm)	Young's modulus (GPa)	Reference
Aspirin	Crystal	–	0.1	Ridgway et al. (1969a)
Aspirin	Compact	+180–250	2.5	Kerridge and Newton (1986)
Aspirin	Compact	+250–355	2.3	Kerridge and Newton (1986)
Sodium chloride	Crystal	–	1.9	Ridgway et al. (1969a)
Potassium chloride	Compact	–	9.2	Kerridge and Newton (1986)
Avicel PH102	Compact	+250–350	4.7	Kerridge and Newton (1986)
Sucrose	Crystal	–	2.2	Ridgway et al. (1969a)
Salicylamide	Crystal	–	1.3	Ridgway et al. (1969a)
Hexamine	Crystal	–	0.9	Ridgway et al. (1969a)

The split Hopkinson bar

In contrast to the other techniques described above, which can only be used to evaluate Young's modulus of elasticity in either tension or compression, the split Hopkinson bar can be used to evaluate both simultaneously on the same specimen.

In the test a flat-faced tablet is sandwiched between two cylindrical steel rods (10 mm diameter) which are aligned horizontally as shown schematically in Fig. 8a (Al-Hassani *et al.*, 1989). The two rods are supported on four adjustable V-slot knife edges which minimize frictional effects and thus prevent disturbance of the stress wave. The porous tablets are held in place using standard strain gauge adhesive which allows the transmission of stresses through the specimen. In the experiment a short compressive loading pulse is initiated by striking the free end of the input bar with a short 10 mm diameter aluminium rod. At the specimen–bar interface the pulse divides into transmitted and reflected components, or compressive and tensile stress waves, respectively. The compressive wave travels to the free end of the output bar, reflecting as a tensile wave to load the specimen

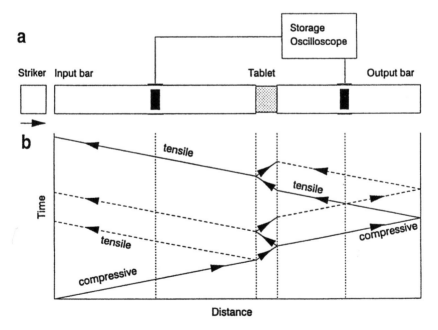

Fig. 8 Schematic diagram of the Split Hopkinson bar with space–time diagram showing pulse partition (adapted from Al-Hassani *et al.*, 1989).

in tension, as shown schematically in Fig. 8b. In time numerous internal reflections occur in the bar and specimen resulting in complex waveforms which are disregarded in the analysis. The stress waves are detected by four strain gauges at each position on the input and output bars, thus maximizing sensitivity and allowing only the measurement of axial strains. The first compressive and tensile incident pulses and their associated reflected and transmitted pulses are used to calculate the elastic constants.

The theory underlying the method is fully described by Al-Hassani *et al.* (1989). Young's modulus either in compression, E_c, or in tension, E_t, can be calculated using the appropriate strain signals using the expression:

$$E_s = \frac{E_R \epsilon_t A l_s}{A_s 2c \int_0^t \epsilon_r \mathrm{dt}} \tag{20}$$

where E_R and E_s are the Young's modulus of elasticity of the bar material and specimen, respectively, A and A_s the cross-sectional area of the bar and specimen, respectively, l_s is the specimen thickness, ϵ_r and ϵ_s are the

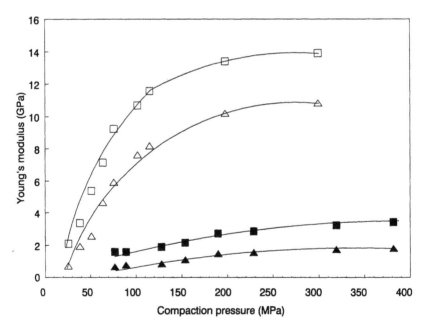

Fig. 9 The effect of compaction pressure on the compressive (squares) and tensile (triangles) Young's modulus for paracetamol DC (filled symbols) and Dipac sugar (unfilled symbols) (taken from Al-Hassani *et al.*, 1989).

reflected and transmitted strain, c is the longitudinal stress wave velocity and $\int_0^t (\epsilon_r) dt$ is the area under the reflected strain curve.

The technique has been used to evaluate the compressive and tensile Young's modulus of two pharmaceutical materials: paracetamol DC (spray granulated paracetamol with 4% hydroysed gelatin) and Dipac sugar (Al-Hassani *et al.*, 1989; Sarumi and Al-Hassani, 1991). Data from these workers are reproduced in Fig. 9. Unfortunately the workers did not report porosity values for the specimens measured and hence the data cannot be evaluated in terms of calculating Young's modulus at zero porosity. However, it can be seen that for all specimens Young's modulus in tension is always lower than in compression.

The tensile modulus (taken as the maximum values reported by Al-Hassani *et al.*, 1989) for Dipac sugar and paracetamol DC are 10.8 and 1.8 GPa, respectively. The value for paracetamol DC is a factor of 10 lower compared with the value measured using three-point beam bending (Roberts, 1991) but the value for Dipac sugar compares favourably with that measured using four-point beam bending (Bassam *et al.*, 1990).

As stated by Al-Hassani *et al.* (1989) the technique has many advantages in that it is simple and precise. For pharmaceutical materials it has the added advantage that the specimens used are 10 mm diameter flat-faced tablets that are easily prepared using standard punches and dies in a tablet machine.

Miscellaneous techniques

For all materials Young's modulus E is directly related to the two other standard moduli – shear modulus, G, and bulk modulus, K, by the equation:

$$E = \frac{9KG}{3K + G} \qquad (21)$$

thus allowing the calculation of any modulus from measurements of the other two. If, however, only one modulus is known the others can be calculated from equations involving a knowledge of the Poisson's ratio, v:

$$E = 3G(1 + v) \qquad (22)$$

$$E = 3K(1 - 2v) \qquad (23)$$

These equations can be used to calculate Young's modulus of elasticity both of crystals from single crystal elastic constants and of powders from viscoelastic properties.

An example of the first approach is the work by Kim *et al.* (1985) who calculated the lattice constants, lattice energy and structural parameters of aspirin crystals by applying lattice dynamics. In this approach the lattice energy is minimized with respect to the lattice parameters, e.g. unit cell constants, by using a model intermolecular potential calculated from the sum of the pairwise atom–atom interaction terms of several types of potential that take account of the electrostatic interactions, van der Waals attractions, a repulsive non-bonded term and hydrogen bonding contributions. The minimum energy structure is then distorted by the application of axial or shear strains and the potential energy of this system is subsequently minimized to obtain strain energy curves and surfaces. Single crystal elastic constants (C_{ij}) can be calculated from the second derivatives of the lattice energy, i.e. of the energy hypersurface minimum. The bulk and shear moduli can then be calculated using Voigt averages:

$$K = (C_{11} + C_{22} + C_{33} + 2C_{12} + 2C_{23} + 2C_{13})/9 \qquad (24)$$

$$G = (C_{11} + C_{22} + C_{33} - C_{12} - C_{23} - C_{13} + 3C_{44} \\ + 3C_{55} + 3C_{66})/15 \qquad (25)$$

where for instance C_{11}, C_{22} and C_{33} are the single crystal elastic constants representing stress and strain in the x-axis, y-axis and z-axis directions and are equivalent to the tensile modulus in those directions, e.g. Young's modulus in the x direction, $E_x = C_{11} = \sigma_x/\epsilon_x$.

Calculation of Young's modulus from equation 21 using data of Kim et al. (1985) gave a value for aspirin of 7.1 GPa which compares favourably with 7.5 GPa obtained from three-point beam bending (Roberts et al., 1991).

An example of the second approach is the work by Radebaugh et al. (1989) who described a method for the determination of the shear modulus of microcrystalline cellulose (Avicel PH102) using a small-strain sinusoidal oscillatory torsion applied to beam specimens of dimensions $122 \times 50 \times 2.8$ mm. Using the data given in the paper and applying a two-term polynomial (Spinner et al., 1963) the shear modulus at zero porosity can be calculated to be 3.5 GPa. If the Poisson's ratio of microcrystalline cellulose is taken as 0.33, as suggested by Church (1984), it is possible to calculate a value for Young's modulus of microcrystalline cellulose of 9.3 GPa. This compares favourably with the value of 8.7 GPa measured using four-point beam bending (Bassam et al., 1990).

INDENTATION HARDNESS AND YIELD STRESS

The hardness and/or yield stress of a material are measures of its resistance to local deformation. For plastic materials the two parameters are directly related, the yield stress being one-third of the hardness. While hardness is generally measured by means of indentation, yield stress may be determined by more indirect methods, e.g. data from compaction studies. Both will be discussed in this section.

Indentation hardness testing

The most common method of measuring the hardness of a material is the indentation method. In this a hard indenter (e.g. diamond, sapphire, quartz, or hardened steel) of specified geometry is pressed into the surface of the material. The hardness is essentially the load divided by the projected area of the indentation to give a measure of the contact pressure. The most commonly used methods for determining the hardness are the Vickers, Brinell, Knoop and Rockwell tests and these are shown in Fig. 10. However, two techniques have been used to measure the yield properties of pharmaceutical materials: the Brinell test and the Vickers test. In general, the former has been used for measurements on compacts whereas the latter technique has been solely applied to single crystals.

In the standard Brinell test (Brinell, 1901) a hard steel ball (usually 1 cm in diameter) is normally pressed on to the surface of the material. The load, F, is applied for a standard period of 30 s and then removed and the diameter, d, of the indentation is measured and the Brinell hardness, H_b, determined by the following relationship:

$$H_b = \frac{F}{(\pi D/2)(D - \sqrt{D^2 - d^2})} \tag{26}$$

where D is the diameter of the ball (Fig. 10).

However, it was found by Brinell that for most materials, the hardness increased as the indent was made larger, this occurring when the chordal diameter is greater than 0.4 of the diameter of the ball. This is caused by work hardening and for these instances the measured hardness is called the Meyer hardness, H_m, and is determined from the load, F, divided by the projected area of the impression:

$$H_m = \frac{4F}{\pi d^2} \tag{27}$$

In the extreme case when the indenter is pressed up to its diameter the Brinell hardness will be one-half the Meyer hardness.

In the Vickers test (Smith and Sandland, 1925) a square-based diamond pyramid is used as an indenter. It is capable of measuring hardnesses over the entire range from the softest to the hardest materials. The Vickers hardness, H_v, is determined from the following equation, where F is the applied force and d is the length of the diagonals of the square impression (Fig. 10):

$$H_v = \frac{2F\sin68^0}{d^2} \qquad (28)$$

The specimen thickness should be at least one and a half times the diagonal length; greatest accuracy is obtained with high loads, but loads as small as 0.01 N can be used. However, at lower loads the elastic recovery is of greater importance. For loads lower than 0.5 N the technique is usually described as microhardness testing. The 136° angle between opposite faces of the Vickers indenter (Fig. 10) is based on geometrical similarity to the Brinell indenter allowing for conversions between hardness scales provided the diameter of the impression is 0.375 times the ball diameter (i.e. the angle included between the tangents of a circle under these conditions for the ball indenter is 136°) and the Brinell hardness < 4000 MPa (Cottrell, 1964).

Indenter	Vickers	Brinell	Knoop	Rockwell
Material of which indenter is made	Diamond	Hardened steel or tungsten carbide	Diamond	Diamond
Shape of indenter	Square based pyramid	Sphere	Rhomb based pyramid	Cone
Dimensions of indenter	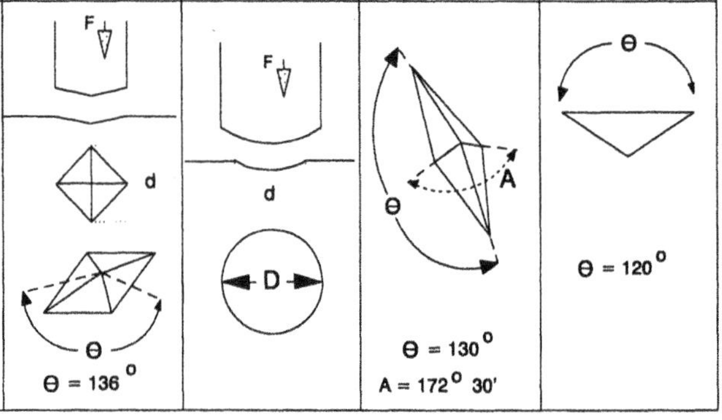 $\Theta = 136^o$	d	$\Theta = 130^o$ $A = 172^o$ 30'	$\Theta = 120^o$

Fig. 10 Geometries for various hardness testers. For definitions of terms, see text.

Although hardness measurements on compacts were first recorded by Spengler and Kaelin (1945) using a Brinell indenter and by Nutter Smith (1949) using a Vickers indenter, the specimens used were formulations containing added excipients and hence are of little use for the evaluation of material properties.

An improved analysis of the Vickers hardness of compacts was performed by Ridgway *et al.* (1969b) using a Leitz microhardness tester with loads of between 5 and 2000 g. Accurate diagonal lengths were determined after lightly dusting the indent with graphite powder. Of the materials studied potassium chloride, hexamine and urea showed little change in hardness with increasing compaction pressure while for aspirin and sodium chloride the hardness increased possibly owing to work hardening. The maximum hardness values reported were 29 MPa, 17 MPa, 16 MPa, 12 MPa and 8 MPa for sodium chloride, potassium chloride, aspirin, hexamine and urea, respectively. These values are an order of magnitude lower than those generally accepted and hence must be viewed with suspicion.

In many studies on compacts, spherical indenters have been used either fitted to commercially available instruments or custom-built equipment. An example of the former is the pneumatic microindentation apparatus described by Ridgway *et al.* (1970). This instrument (Plate 4) can apply loads of between 4 and 8 g using a spherical indenter 1.5 mm in diameter and measure depths of penetration of 1–6 μm. Using this apparatus on aspirin compacts, the authors were able to show that hardness measurements at the centre of compacts were higher than those at the periphery, a property for flat-faced compacts confirmed by Aulton and Tebby (1975). However, for compacts prepared using concave punches hardness distribution tends to vary with the degree of curvature (Aulton and Tebby, 1975).

In a modification to the original microindentation apparatus (Ridgway *et al.*, 1970), Aulton *et al.* (1974) added a displacement transducer to measure the vertical displacement. Compacts were formed under a compression pressure of 50 MPa and therefore would be expected to have high porosities. They suggested that the elastic quotient index, which is the fraction of the indentation that rebounds elastically, was a measure of the ability of materials to form tablets.

In a further study involving a larger range of materials, Aulton (1981) found that hardness measurements were generally higher on the upper face of the compact than on the lower. However, the differences were material dependent and the hardness measurements reported were 62 MPa, 54 MPa, 51 MPa, 36 MPa, 19 MPa and 13 MPa for sucrose, Sta-Rx, Emcompress, Avicel PH101, lactose β anhydrous and paracetamol, respectively. As well as being extremely low, the relative order of the materials is considered to be wrong in relation to the particle hardness (see later).

Probably the most relevant of all the work carried out on hardness measurement on compacts is that of Hiestand *et al.* (1971) using a spherical indenter attached to a pendulum and that of Leuenberger (1982) and his co-workers (Jetzer *et al.*, 1983a,b, 1985; Leuenberger and Rohera, 1985; Galli and Leuenberger, 1986) using a spherical indenter attached to a universal testing instrument.

In the dynamic pendulum method of Hiestand *et al.* (1971) a 24.5 mm diameter spherical ball falls under the influence of gravity and the rebound height and indent dimensions are measured (Fig. 11). The hardness is calculated from the expression:

$$H_b = \frac{4mgrh_r}{\pi a^4}\left(\frac{h_i}{h_r} - \frac{3}{8}\right) \qquad (29)$$

where m is the mass of the indenter, g is the gravitational constant, r is the radius of the sphere, a is the chordal radius of the indent, h_i is the initial height of the indenter and h_r is the rebound height of the indenter.

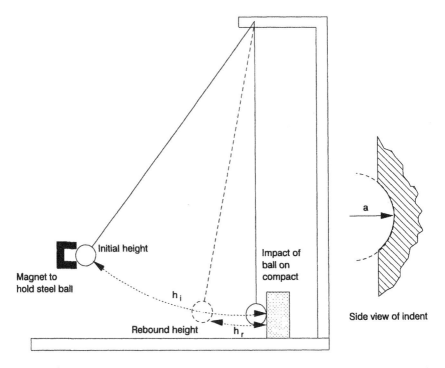

Fig. 11 Schematic diagram of the dynamic pendulum apparatus used by Hiestand *et al.* (1971). For definitions of terms, see text.

In the case of the indenter attached to the universal testing instrument (Leuenberger, 1982) loads of 3.92 and 9.81 N were applied to a sphere of 1.76 mm diameter at a velocity of 0.05 cm min^{-1}; indent diameters were determined from scanning photomicrographs.

In both cases indentation hardness was found to be dependent on the compaction pressure and hence relative density (porosity) of the compact in an exponential manner. While Hiestand *et al.* (1971) considered extrapolation to unit relative density to be questionable, Leuenberger (1982) used the relationship between Brinell hardness, H_b, of a compact and its relative density, D, to develop a measure of compactibility and compressibility, i.e.:

$$H_b = H_{b_{max}}[1 - \exp^{(-\lambda \sigma_c D)}] \qquad (30)$$

where $H_{b_{max}}$ is the theoretical maximum hardness as the compressive stress, σ_c, approaches infinity and the relative density of the compact approaches unity and λ is the rate at which H_b increases with increasing compressive stress. In the equation Leuenberger (1982) has suggested that $H_{b_{max}}$ describes the compactibility and λ compressibility.

Data on indentation hardness using both methods are shown in Table 4. As with the modulus of elasticity, values for hardness vary over two orders of magnitude from the very hard materials (e.g. Emcompress) to very soft waxes. Drugs have intermediate hardness values.

It should be noted that the values recorded are also variable owing to:

1. The intrinsic variability in the specimens. Jetzer *et al.* (1983a) demonstrated that for aspirin, Avicel PH101, caffeine, α-lactose monohydrate and Emcompress the 95% confidence limits gave 82–101 MPa, 148–189 MPa, 163–416 MPa, 409–659 MPa and 210–1294 MPa, respectively. The increased variability also mirrors the increase in hardness (Table 4) for this series of materials.

2. Work hardening as the compaction pressure is increased. Leuenberger and Rohera (1986) and Aulton and Marok (1981) have clearly demonstrated an increase in the Meyers work-hardening index for a variety of materials.

3. The increase in hardness with increasing indentation load (Leuenberger and Rohera, 1986).

4. The rate of measurement. In a comparison of the dynamic pendulum method of Hiestand *et al.* (1971) – equivalent to a strain rate equal to that of a high-speed compaction simulator – and the pseudostatic method of Leuenberger (1982), Jetzer *et al.* (1985) demonstrated that, in general, both methods gave the same hardness for aspirin, caffeine and mannitol. However, the dynamic method was less reproducible.

Table 4 Indentation hardness measured on compacts

Material	Indentation hardness (MPa)	Reference
Sugars		
α-Lactose monohydrate	515	Leuenberger (1982)
α-Lactose monohydrate	534	Jetzer *et al.* (1983a)
Lactose β anhydrous	251	Leuenberger (1982)
Sucrose	1046–1723	Leuenberger (1982)
Sucrose (250–355#)	493	Jetzer *et al.* (1983a)
Drugs		
Paracetamol DC	265	Jetzer *et al.* (1983a)
Caffeine (anhydrous)	290	Jetzer *et al.* (1983a)
Caffeine (granulate)	288	Jetzer *et al.* (1983a)
Oxprenolol succinate	262	Galli and Leuenberger (1986)
Hexamine	232	Leuenberger (1982)
Phenacetin	213	Leuenberger (1982)
Sitosterin	198	Leuenberger (1982)
Metamizol	91	Jetzer *et al.* (1983a)
Aspirin powder	91	Jetzer *et al.* (1983a)
Aspirin FC	87	Jetzer *et al.* (1983a)
Aspirin	60	Leuenberger (1982)
Aspirin	55	Leuenberger (1982)
Ibuprofen (A)	35	Leuenberger (1982)
Ibuprofen (B)	162	Leuenberger (1982)
Alkali halides		
NaCl	653	Leuenberger (1982)
NaCl	313	Jetzer *et al.* (1983a)
NaCl (rock salt)	358	Jetzer *et al.* (1983a)
KCl	99	Jetzer *et al.* (1983b)
KBr	69	Jetzer *et al.* (1983a)
Others		
Emcompress	752	Jetzer *et al.* (1983a)
Avicel PH102	168	Jetzer *et al.* (1983a)
Starch 1500	78	Galli and Leuenberger (1986)
Sodium stearate	37	Leuenberger and Rohera (1985)
PEG 4000	36	Leuenberger and Rohera (1985)
Castor oil (hydrogenated)	32	Galli and Leuenberger (1986)
Magnesium stearate	22	Leuenberger and Rohera (1985)
Sodium lauryl sulfate	10	Leuenberger and Rohera (1985)

PEG, polyethylene glycol.

Hardness measurements on single crystals of pharmaceutical materials are sparse but where performed all have been carried out using a pyramidal indenter.

The earliest reported study on the hardness of pharmaceutical crystals was carried out by Ridgway *et al.* (1969a) using the Leitz microhardness tester with loads of 25 g or less. The crystals were mounted in heat-softened picene wax on a mounting slide to ensure that the surfaces were horizontal. It is interesting to note that the authors observed that aspirin and sucrose showed cracking and regarded this as a problem with the technique. Hardness values from this study are presented in Table 5. It is interesting to note that softer materials showed the most variation in results owing to a decline in the definition and quality of the indent.

The next reported study of crystal hardness using a Vickers hardness

Table 5 Indentation hardness measured on crystals

Material	Indentation hardness (MPa)	Reference
Sugars		
α-Lactose monohydrate	523	Ichikawa *et al.* (1988)
Sucrose	645	Duncan-Hewitt and Weatherly (1989b)
Sucrose	636	Ridgway *et al.* (1969a)
Drugs		
Paracetamol	421	Duncan-Hewitt and Weatherly (1989b)
Paracetamol	342	Ichikawa *et al.* (1988)
Sulfaphenazole	289	Ichikawa *et al.* (1988)
Hexamine	133	Ridgway *et al.* (1969a)
Hexamine	42	Ichikawa *et al.* (1988)
Sulfadimethoxine	231	Ichikawa *et al.* (1988)
Phenacetin	172	Ichikawa *et al.* (1988)
Salicylamide	151	Ridgway *et al.* (1969a)
Salicylamide	123	Ichikawa *et al.* (1988)
Aspirin	87	Ridgway *et al.* (1969a)
Alkali halides		
NaCl	212	Ridgway *et al.* (1969a)
NaCl	213	Duncan-Hewitt and Weatherly (1989b)
NaCl	183	Ichikawa *et al.* (1988)
KCl	177	Ridgway *et al.* (1969a)
KCl	101	Ichikawa *et al.* (1988)
Others		
Urea	91	Ridgway *et al.* (1969a)
Urea	83	Ichikawa *et al.* (1988)

tester was carried out 17 years later by Ichikawa *et al.* (1988). In this study the majority of the materials (with the exception of sucrose and urea) were recrystallized. Indentation was performed on the crystal face possessing the largest area, i.e. the face that grows the slowest during crystallization, as it was inferred that this face would have the greatest influence on the compaction properties. Ichikawa *et al.* (1988) attributed differences in crystal hardness as a reflection of the mechanism of deformation during compaction. They also showed that the reciprocal of crystal hardness correlated with the slope K from the Heckel equation (see later). Data reproduced in Table 5 represent the mean value from three different applied loads: 10 g, 25 g, 50 g. The importance of load on the indentation test should be noted as in the case of crystals the hardness generally decreases as load is increased and this is the exact opposite to that seen for compacts.

More recently Duncan-Hewitt and Weatherly (1989a, 1989b) measured the Vickers hardness on single crystals of a number of materials using a Leitz–Wetzlar Miniload tester with a load of 147 mN. It is interesting to note that the surfaces of sucrose were preconditioned, by washing with methanol then polishing by abrading with decreasing grades of emery paper. This may account for their higher hardness values compared with compacts (Table 4) as in this case the materials could be fully work-hardened solids. However, the authors did not indicate whether the other crystals were preconditioned. Furthermore Duncan-Hewitt and Weatherly (1989a) showed that the sucrose crystals were anisotropic in that different crystal faces gave different hardness values (Table 6) where the (001) and (100) are the predominant faces (i.e., have the largest area).

Finally, a brief mention must be made of the effects of dislocations on hardness, as crystallization can affect the number and distribution of these defects and therefore affect the resultant mechanical properties. In general, increasing the crystallization rate tends to increase the number of dislocations and therefore increase the indentation hardness. Such effects have

Table 6 The effect of crystal face on the indentation hardness of sucrose (Duncan–Hewitt and Weatherly, 1989a)

Crystal face	Hardness (MPa)
001	636
100	649
110	642
010	649

been noted by Hiestand *et al.* (1981) and Wong and Aulton (1987) for ibuprofen and α-lactose monohydrate respectively.

Yield stress from compaction studies

Compaction studies, because they mimic the tabletting process, offer an ideal method for assessing the mechanical properties of powders. In powder compaction a specific method used to evaluate the average stress of a material during compression relies on the observations of Heckel (1961a,b) who found that for materials that plastically deform, the relative density of a material, D, could be related to the compaction pressure, P, by the equation:

$$\ln\left[\frac{1}{(1-D)}\right] = KP + A \qquad (31)$$

where K and A are constants.

Unfortunately, considerable deviations of the experimental data occur at both low and high pressures owing to particle rearrangement and strain hardening, respectively, but at least over the middle pressure range a straight line relationship exists between $\ln(1/1 - D)$ and P (Fig. 12).

Equation 31 has been reappraised by Hersey and Rees (1970) who suggested that the reciprocal of K can be regarded as numerically equal to the

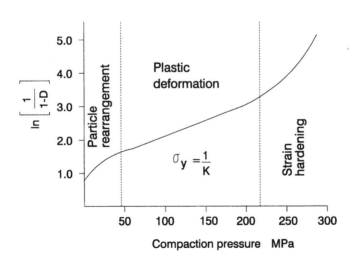

Fig. 12 Schematic diagram of the Heckel plot (Heckel, 1961a,b). For definitions of terms, see text.

mean yield stress of the powder. However, as pointed out by Roberts *et al.*
(1989a) this is only a specific case and the reciprocal of K can be regarded
as a mean deformation stress, be it a plastic deformation stress (equal to
the yield stress) for materials that deform plastically, a fracture deformation stress for materials that undergo fracture or a combination of the two.
This approach implies that provided the experiment is carried out on
materials close to or below the brittle–ductile transition (Fig. 1) then the
reciprocal of K will be numerically equal to the yield stress of the powder.

Historically, two approaches to the analysis of Heckel data have been
used and these are generally referred to as 'at pressure' and 'zero pressure'
measurements. Pressure/relative density measurements determined during
compression are clearly 'at pressure' measurements while those from
relative density measurements on the compact after ejection are 'zero pressure' measurements. In the original publication Heckel (1961a) found that
for the metals iron, copper, nickel, and tungsten there was no difference
between the two measurements but for graphite the 'zero pressure' measurement was higher and could be attributed to the elastic recovery of the compact causing a lower relative density. Support for this hypothesis can be
obtained from studies on the pharmaceutical materials dicalcium phosphate
dihydrate (8% increase, Paronen, 1987), lactose (30% increase, Fell and
Newton, 1971), microcrystalline cellulose (56% increase, Paronen, 1987)
and starch (177% increase, Paronen, 1987) where the magnitude of the difference is indirectly related to the modulus of the material, i.e. the larger
the increase the lower the modulus.

In the light of the discussion above and the findings that other factors
such as punch and die dimensions (York, 1979; Danjo *et al.*, 1989), state
of lubrication (DeBoer *et al.*, 1978; Ragnarsson and Sjogren, 1984), and
speed of compaction (Rees and Rue, 1978; Roberts and Rowe, 1985) can
have an effect on the measurement, it is not surprising that a great deal
of controversy and confusion surrounds the use of data from Heckel plots.
However, Roberts and Rowe (1987a) have clearly shown that, provided
measurements are carried out 'at pressure' with lubricated punches and dies
on material that is below its brittle–ductile transition and at a very slow
speed, the reciprocal of K is identical to the yield stress of the material and
comparable with that calculated from indentation hardness measurements
using equation 5 (Table 7).

Although early measurements were generally performed on either instrumented punches and dies in physical testing machines (Fell and Newton,
1971) or instrumented single-punch tablet machines (DeBoer *et al.*, 1978)
with relatively unsophisticated data capture and analysis, recent measurements generally have been carried out using tablet compression simulators.

Table 7 Yield stresses from the Heckel plot compared with those determined by the indentation technique

Material	Yield stress (MPa)	
	Heckel plot	Indentation
Copper	331	390
Lactose	174	178
Mannitol	90	103
NaCl	89	86
Avicel	48	56
PTFE	12	26

PTFE, polytetrafluoroethylene.

One such simulator, as used by Roberts and Rowe (1985), is shown in Plate 5. It consists of three units: a hydraulic power pack (A), an electronic control unit (B) and a loading frame (C). The hydraulic system consists of a continuously circulating closed loop of hydraulic fluid supplying three servo valves, two in parallel on the upper actuator and one on the lower. These receive drive signals from the electronic control unit and the profile generator. The two actuators can independently move the two rams and can hold both single punch and rotary punch tooling with conventional dies with a total stroke of 50 mm and 25 mm for the upper and lower rams, respectively. The actuators can achieve compression forces up to 50 kN during a compression cycle in any time interval between 40 ms and 1000 s controlled by a clock generator. The punches are capable of a maximum punch velocity of 300 mm s^{-1} using a sawtooth displacement–time profile with a displacement accuracy of ±10 μm and a force accuracy of ±0.025 kN.

Yield stress data for a number of excipients and drugs are shown in Table 8. The trends seen mimic those for indentation hardness with the inorganic carbonates and phosphates showing very high values of yield compared with the polymers, which give low yield values. The variation seen in the drugs may well be due to slight differences in the particle sizes tested.

Compaction simulators allow the measurement of yield stress over a wide range of punch velocities (Fig. 13). To compare materials, Roberts and Rowe (1985) proposed the term 'strain rate sensitivity' (SRS) to describe the percentage decrease in yield stress from a punch velocity of 300 mm s^{-1} to one of 0.033 mm s^{-1}. This was later modified (Roberts and Rowe, 1987a) to a percentage increase in yield stress over the same punch velocities:

Table 8 Yield stresses measured by compaction studies

Material	Experimental details	Yield stress (MPa)	References
Inorganic			
Calcium phosphate	Simulator	957	Roberts and Rowe (1987a)
Calcium carbonate	Simulator	851	Roberts and Rowe (1987a)
Calcium carbonate	Hydraulic press	610	Ejiofer et al. (1986)
Magnesium carbonate	Simulator	471	Roberts and Rowe (1987a)
Dicalcium phosphate dihydrate	Simulator	431	Roberts and Rowe (1987a)
Sugars			
α-Lactose monohydrate	Hydraulic press	179	Vromans and Lerk (1988)
α-Lactose monohydrate	Hydraulic press	183	York (1978)
α-Lactose monohydrate	Simulator	178	Roberts and Rowe (1986)
Lactose β anhydrous	Simulator	149	Roberts and Rowe (1987a)
Lactose (spray dried)	Simulator	178	Bateman et al. (1989)
Lactose (spray dried)	Simulator	147	Roberts and Rowe (1985)
Mannitol	Simulator	90	Roberts and Rowe (1987a)
Others			
Sodium chloride	Single punch	89	Ragnarsson and Sjogren (1985)
Sodium chloride	Simulator	89	Roberts et al. (1989a)
Avicel PH101	Single punch	50	Humbert-Droz et al. (1982)
Avicel PH101	Simulator	46	Roberts and Rowe (1986)
Avicel PH102	Simulator	49	Roberts and Rowe (1986)
Avicel PH105	Simulator	48	Roberts and Rowe (1986)
Maize starch	Simulator	40	Roberts and Rowe (1987a)
Stearic acid	Hydraulic press	4.5	York (1978)

Polymers

PVC/vinyl acetate	Simulator	70	Roberts and Rowe (1987a)
Polyethylene	Simulator	16	Roberts and Rowe (1987a)
PTFE	Simulator	12	Roberts and Rowe (1987a)

Drugs

Paracetamol DC	Single punch	108	Hussain et al. (1991)
Paracetamol DC	Single punch	81	Humbert-Droz et al. (1982)
Paracetamol DC	Simulator	109	Roberts and Rowe (1987a)
Paracetamol	Single punch	79	Humbert-Droz et al. (1983)
Paracetamol	Single punch	99	Podczeck and Wenzel (1989)
Paracetamol	Single punch	127	Duberg and Nystrom (1986)
Paracetamol	Simulator	102	Roberts and Rowe (1985)
Sulphathiazole	Hydraulic press	109	Ramberger and Burger (1985)
Theophylline (anhydrous)	Single punch	75	Podczeck and Wenzel (1989)
Aspirin	Single punch	25	Humbert-Droz et al. (1983)
Aspirin	Single punch	73	Duberg and Nystrom (1986)
Tolbutamide	Single punch	24	Humbert-Droz et al. (1983)
Phenylbutazone	Single punch	24	Humbert-Droz et al. (1983)
Ibuprofen	Simulator	25	Bateman et al. (1987)

PTFE, polytetrafluoroethylene; PVC, polyvinyl chloride.

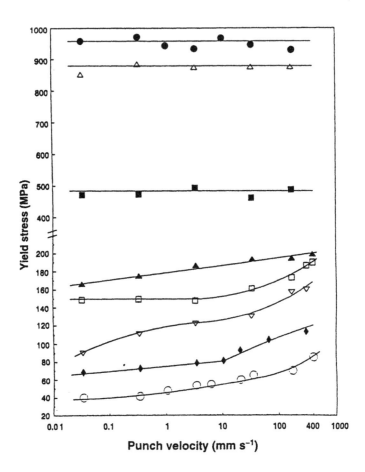

Fig. 13 The effect of punch velocity on yield stress for ●, calcium phosphate; △, calcium carbonate; ■, magnesium carbonate; ▲, α lactose monohydrate; □, lactose β anhydrous; ▽, mannitol; ◆, Avicel PH101; ○, maize starch.

$$\text{SRS} = \frac{\sigma_{y300} - \sigma_{y0.033}}{\sigma_{y0.033}} \tag{32}$$

where σ_{y300} and $\sigma_{y0.033}$ are the yield stresses measured at punch velocities of 300 and 0.033 mm s^{-1}, respectively.

Data on a number of materials are shown in Table 9. Some materials such as the inorganic carbonates/phosphates show little rate dependence while others such as starch and mannitol show a large strain rate dependence

Table 9 Strain rate sensitivities (SRS) of some excipients and drugs

Material	SRS (%)
Calcium phosphate	0
Calcium carbonate	0
Heavy magnesium carbonate	0
Paracetamol DC	1.8
Paracetamol	11.9
α-Lactose monohydrate (fine grade)	19.4
Lactose (spray dried)	23.8
Lactose β anhydrous	25.5
Avicel PH101	63.7
Sodium chloride	66.3
PVC/vinyl acetate	67.5
Mannitol	86.5
Maize starch	97.2

PVC, polyvinyl chloride.

consistent with differences in the time-dependent properties of the material during compaction.

The effects of moisture and particle size of materials will be discussed later in this chapter.

CRITICAL STRESS INTENSITY FACTOR, K_{IC}

The critical stress intensity factor, K_{IC}, describes the state of stress around an unstable crack or flaw in a material and is an indication of the stress required to produce catastrophic propagation of the crack. It is thus a measure of the resistance of a material to cracking. Since it is related to the stress and the square root of crack length it has the dimensions of $MPam^{1/2}$.

All methods used to measure K_{IC} involve specimens containing induced notches and/or cracks and for pharmaceutical materials include three- or four-point beam bending (commonly known as the single edge notched beam test or SENB), double torsion, radial edge cracked tablet or disc and Vickers indentation. The choice of the test and its associated specimen geometry depends on the rate of testing, the ease of formation of the specimen and the porosity of the specimen. Specimen porosity is specifically important since pores can act as stress concentrators. However in all testing methods the influence of pores is minimized by the induction of a dominant crack or flaw often in the form of a notch.

Single edge notched beam

In this test a prenotched rectangular beam of small thickness and width in comparison to its length is subjected to transverse loads and the load at fracture measured. As with the beams used for the determination of Young's modulus, loading can be by either three or four points (Fig. 14). The single edge notched beam test has been the subject of much research leading to the specification of standard criteria for test piece geometry (Brown and Srawley, 1966; British Standards Institution, 1977) dependent on the material to be studied, e.g.:

$$c \text{ (crack length)} > 2.5 \left(\frac{K_{IC}}{\sigma_y}\right)^2 \tag{33}$$

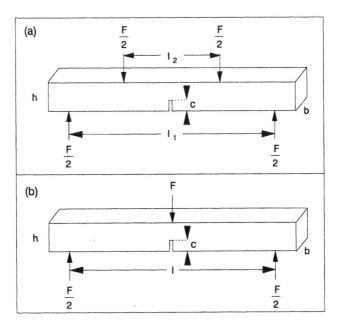

Fig. 14 Geometries for (a) four-point and (b) three-point single edge notched beam. F, applied load; h, beam thickness; b, beam width; l and a, distances between loading points as shown; c, crack length.

$$h \text{ (specimen height)} > 2.5 \left(\frac{K_{IC}}{\sigma_y}\right)^2 \tag{34}$$

To take account of the uncertainties such as underestimation of the K_{IC} value and the possibility of the test not meeting some of the other validity criteria set down in British Standard 5447 (1977) it is recommended that larger pieces be used where c and h should be at least $4\left(\frac{K_{IC}}{\sigma_y}\right)^2$. In addition, Brown and Srawley (1966) have recommended that:

$$w \text{ (specimen width)} > 5.0 \left(\frac{K_{IC}}{\sigma_y}\right)^2 \tag{35}$$

In all cases σ_y is the yield stress of the material under test.

Equations for the calculation of the critical stress intensity factor from the applied load, F, and geometry of the beam can be derived, e.g. for four-point beam bending (Mashadi and Newton, 1987a,b):

$$K_{IC} = \gamma \frac{3Fc^{1/2}(l_1 - l_2)}{2bh^2} \tag{36}$$

and for three-point beam bending (Roberts et al., 1993):

$$K_{IC} = \gamma \frac{3Fc^{1/2}}{2bh^2l} \tag{37}$$

where γ is a function of the specimen geometry expressed as a polynomial of the parameter c/h:

$$\gamma = A_0 + A_1\left(\frac{c}{h}\right) + A_2\left(\frac{c}{h}\right)^2 + A_3\left(\frac{c}{h}\right)^3 + A_4\left(\frac{c}{h}\right)^4 \ldots \tag{38}$$

where the coefficients have the values shown in Table 10. These equations assume that the artificially induced crack which becomes unstable has zero width, extends the full width/breadth of the specimen and has a depth that is precisely known. Furthermore, the cross-section of the specimen and of the notch must be of sufficient size relative to the microstructural features that the observed response to loading is representative of the bulk material.

The four-point single edge notched beam was first used for pharmaceutical materials by Mashadi and Newton (1987a,b) and has since been adopted by York et al. (1990). In all cases large rectangular beams 100 mm long × 10 mm wide of varying height are prepared using the same punches and dies as those used to prepare specimens for the determination of Young's modulus. Notches of varying dimensions and profiles have been

Table 10 Values of coefficient A as used in equation 38

Type of loading	A_0	A_1	A_2	A_3	A_4
Four point	+1.99	−2.47	+12.97	−23.17	+24.80
Three point					
$\frac{c}{h} = 8$	+1.96	−2.75	+13.66	−23.98	+25.22
$\frac{c}{h} = 4$	+1.93	−3.07	+14.53	−25.11	+25.80

introduced by cutting either using a simple glass cutter (Mashadi and Newton, 1987a,b) or a cutting tool fitted into a lathe (York *et al.*, 1990). The latter method has allowed notches of different profiles and dimensions to be accurately cut and investigated. While the arrowhead type notch did appear to influence the measured value of K_{IC} for beams of microcrystalline cellulose, the effect was much reduced from straight-through notches and hence the latter were recommended (York *et al.*, 1990).

The load required for failure of the specimens under tension is measured using the same testing rig as that described previously for the determination of Young's modulus. Loading rates of between 0.025 mm min^{-1} (Mashadi and Newton, 1987a,b) and 100 mm min^{-1} (York *et al.*, 1990) have been used – the latter workers noted a small rise (approximately 10%) in the measured K_{IC} of beams of microcrystalline cellulose for a 100-fold increase in applied loading rate.

As with the measurement of Young's modulus the four-point test requires large beams and consequently large amounts of material. In order to minimize the latter, especially for the measurement of the critical stress intensity factor for drugs under development, Roberts *et al.* (1993) have developed a three-point test using specimen dimensions and testing rig (Plate 6) similar to that described earlier for Young's modulus although in this case a tensometer has been used to stress the specimen. Using beams of dimensions 20 mm long × 7 mm wide of varying height with two types of notches (a V notch cut by a razor blade pressed into the surface and a straight-through notch cut by a small saw blade) the authors were able to show equivalence with data generated by York *et al.* (1990) for the four-point single edge notched beam (Fig. 14).

As can be seen from Fig. 15, specimen porosity has a significant effect on the measured K_{IC}. As porosity decreases K_{IC} increases indicating more resistance to crack propagation. During the initial stages of compression the large pores which control the strength of the specimen are removed first followed by the smaller ones. As the powder becomes more consolidated it becomes less brittle and is able to absorb greater loads before failure. It

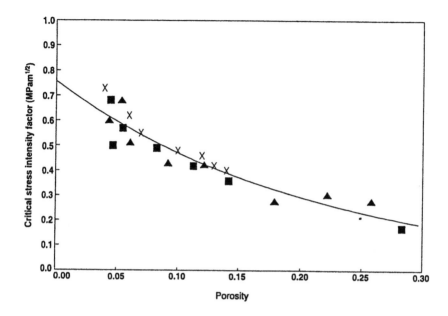

Fig. 15 The effect of porosity on the measured critical stress intensity factor for: ■, V notch; and ▲, straight through notch of Avicel PH101 under three-point loading. X, data from York *et al.* (1990) for some material under four-point loading. The line represents equation 40 for all points.

is obvious from the results in Fig. 15 that the relationship between K_{IC} and porosity is not linear as suggested by Mashadi and Newton (1987a,1987b, 1988). In this respect York *et al.* (1990) have investigated the application of both the two-term polynomial equation of the type:

$$K_{IC} = K_{ICo}(1 - f_1 P + f_2 P^2) \qquad (39)$$

and the exponential equation of the type:

$$K_{IC} = K_{ICo}\exp^{-bP} \qquad (40)$$

where K_{ICo} is the critical stress intensity factor at zero porosity, K_{IC} is the measured critical stress intensity factor of the specimen at porosity P and b, f_1, and f_2 are constants. (It should be noted that these equations are analogous to those used previously for Young's modulus.) Both the relationships give low standard errors and high correlation coefficients for microcrystalline cellulose from various sources. As with Young's modulus, the exponential relationship is the preferred option for pharmaceutical materials (Roberts *et al.*, 1993).

Table 11 Critical stress intensity factors measured using a single-edge notched beam

Material	Method	Critical stress intensity factor ($MPam^{1/2}$)	Particle size (μm)	Reference
Celluloses				
Avicel PH101	4PB (100 × 10) LIN	1.21	50	Mashadi and Newton (1987b)
Avicel PH102	4PB (100 × 10) EXP	0.76	90	York *et al.* (1990)
Avicel PH101	4PB (100 × 10) EXP	0.87	50	York *et al.* (1990)
Avicel PH105	4PB (100 × 10) EXP	1.33	20	York *et al.* (1990)
Emcocel (90M)	4PB (100 × 10) EXP	0.80	90	York *et al.* (1990)
Emcocel	4PB (100 × 10) EXP	0.92	56	York *et al.* (1990)
Unimac (MG200)	4PB (100 × 10) EXP	0.67	103	York *et al.* (1990)
Unimac (MG100)	4PB (100 × 10) EXP	0.80	38	York *et al.* (1990)
Avicel PH102	4PB (100 × 10) POLY	0.91	90	York *et al.* (1990)
Avicel PH101	4PB (100 × 10) POLY	0.99	50	York *et al.* (1990)
Avicel PH105	4PB (100 × 10) POLY	1.42	20	York *et al.* (1990)
Emcocel (90M)	4PB (100 × 10) POLY	0.83	90	York *et al.* (1990)
Emcocel	4PB (100 × 10) POLY	0.80	56	York *et al.* (1990)
Unimac (MG200)	4PB (100 × 10) POLY	0.76	103	York *et al.* (1990)
Unimac (MG100)	4PB (100 × 10) POLY	1.05	38	York *et al.* (1990)
Avicel PH101	3PB (20 × 7) EXP	0.76	50	Roberts *et al.* (1992)

Sugars				
Lactose β anhydrous	3PB (20 × 7) EXP	0.76	149	Roberts *et al.* (1993)
α-Lactose monohydrate	3PB (20 × 7) EXP	0.35	20	Roberts *et al.* (1993)
Sucrose	3PB (20 × 7) EXP	0.22	74	Roberts *et al.* (1993)
Sorbitol instant	4PB (100 × 10) LIN	0.47	–	Mashadi and Newton (1987a)
Drugs				
Ibuprofen	3PB (20 × 7) EXP	0.10	47	Roberts *et al.* (1993)
Aspirin	3PB (20 × 7) EXP	0.16	32	Roberts *et al.* (1993)
Paracetamol DC	3PB (20 × 7) EXP	0.25	120	Roberts *et al.* (1993)
Paracetamol	3PB (20 × 7) EXP	0.12	15	Roberts *et al.* (1993)
Others				
Sodium chloride	3PB (20 × 7) EXP	0.48	20	Roberts *et al.* (1993)
Adipic acid	3PB (20 × 7) EXP	0.14	176	Roberts *et al.* (1993)

4PB = four-point beam; 3PB = three-point beam; EXP = equation 40; POLY = equation 39; LIN = linear regression; the beam dimensions (length and width) are given in parentheses.

A number of pharmaceutical materials have been measured using both three- and four-point beam bending (Table 11). Of all the excipients tested microcrystalline celluloses exhibited the highest values of K_{ICo}; real differences existed between the materials obtained from different sources. In addition, there is also a particle size effect in that, for each of the three sources of material, the critical stress intensity factor increased with decreasing particle size. The sugars exhibit intermediate values with relatively low values for the drugs. It is interesting that anhydrous β lactose has a critical stress intensity factor approximately twice that of α-lactose monohydrate (c.f. indentation hardness measurements in Table 4).

Double-torsion testing

A specific problem with the single edge notched beam test is the process of prenotching the specimen before testing. In addition the procedure has been criticized by Evans (1974) as not being entirely satisfactory for porous specimens. For such specimens Evans (1974) considered the double-torsion method first derived by Outwater and Jerry (1966) and developed by Kies and Clark (1969) to be more appropriate as it eliminates the need to measure crack length.

The specimen is a rectangular plate (Fig. 16) with a narrow groove extending its full length supported on four hemispheres. The load is applied by two hemispheres attached to the upper platten. Controlled precracks are introduced in the specimen by preloading until a 'pop in' is observed (a pop in is a momentary decrease in load and is an indication of crack growth). It should be noted that the groove is necessary to help guide the crack and ensure it remains confined within the groove itself. The critical stress intensity factor is then calculated from the load F required to cause catastrophic crack propagation leading to failure by the expression:

$$K_{IC} = FW_n \left[\frac{3(l + v)}{Wh^3 h_n} \right]^{1/2} \tag{41}$$

where v is the Poisson's ratio and l, h, h_n, W, and W_n are the dimensions of the specimen given in Fig. 16.

The double-torsion method has been used only for microcrystalline cellulose (Avicel PH101) and Sorbitol 'instant' by Mashadi and Newton (1988). As with the single edge notched beam specimens, measurements varied with specimen porosity and using linear extrapolation Mashadi and Newton (1988) calculated values of K_{ICo} of 1.81 and 0.69 MPam$^{1/2}$ for the two materials, respectively. The higher results obtained for these two materials compared with the data obtained for single edge notched beams (1.21

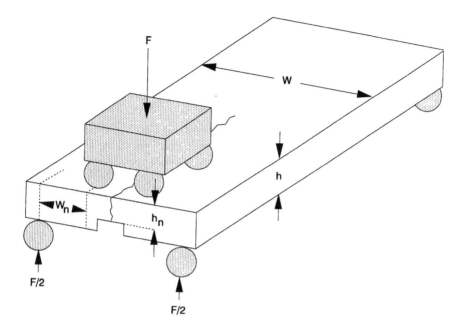

Fig. 16 Geometry for the double torsion method for measuring critical stress intensity factor. *F*, applied load; *h*, plate thickness; *W*, plate width; h_n and W_n, distances as shown.

and 0.47 MPam$^{1/2}$ respectively) have been explained in terms of the specific geometries and stress uniformity of the two techniques. However, it is known that values of K_{IC} determined from double-torsion techniques are generally greater than those from notched beam specimens (Evans, 1974).

A specific practical problem of the double-torsion method for pharmaceutical materials is the preparation of the specimen and the very large compaction pressures needed to produce specimens of low enough porosity. For microcrystalline cellulose, Mashadi and Newton (1988) were only able to produce specimens of greater than 25% porosity.

Radial edge cracked tablets

All the techniques for the determination of the critical stress intensity factor so far described involve the preparation of compressed rectangular beams or plates that require special punches and dies and in some cases high tonnage presses. The ideal specimen shape for pharmaceutical materials is the right-angled cylinder or a flat-faced tablet. Such a shape has recently been

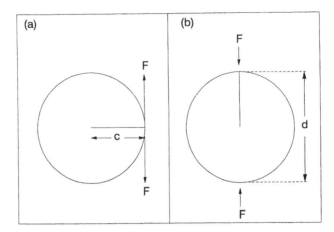

Fig. 17 Geometries for (a) edge opening and (b) diametral compression method for measuring critical stress intensity factors for radially edge cracked tablets. F, applied load; d, tablet diameter; c, crack length.

investigated by Kendall and Gregory (1987) who reported three methods of determining the critical stress intensity factor of precracked specimens, i.e. (a) edge opening (Fig. 17a), (b) diametral compression (Fig. 17b) and (c) pin loading. They concluded that while the first of these was the best test, all had advantages over other test procedures because of their exact theory, simple sample preparation, ease of precracking, straight-forward loading and low propagation forces. The first two methods have recently been evaluated on tablets of microcrystalline cellulose (Avicel PH101) by Roberts and Rowe (1989).

Both the tests involve the introduction of a precrack into the edge of the disc. In edge opening the critical stress intensity factor is given by (Kendall and Gregory, 1987):

$$K_{IC} = \frac{F}{t}\left[\frac{c}{0.3557(d-c)^{3/2}} + \frac{2}{0.9665(d-c)^{1/2}}\right]\left[\frac{c}{2d}\right]^{-1/2} \qquad (42)$$

where F is the peak load for cracking, as in Fig. 17a, while for diametral compression the critical stress intensity factor is given by (Kendall and Gregory, 1987):

$$K_{IC} = \frac{F}{dt}\left[\frac{\pi}{2c}\right]^{-1/2}\frac{1.586}{[1-(c/d)]^{3/2}} \qquad (43)$$

where F is the compressive force for cracking (Fig. 17b). In both cases c is the crack length, t is the tablet thickness and d is the tablet diameter.

When testing microcrystalline cellulose (Avicel PH101) Roberts and Rowe (1989) first prepared tablets of varying porosity using 15 mm flat-faced punches on an instrumented tablet press. After precracking using a scalpel blade (Plate 7) the cracked tablets were stressed using either edge opening, in which tablets were gripped using the air jaws of a tensometer and pulled apart (Plate 8), or diametrally compressed with the crack vertical between the plattens of a tensometer. During extensive testing it was concluded that, of the two techniques, edge opening was the preferred option as it gave the most stable crack propagation and the effects of crack length were minimal. However, the diametral compression test was found to have the same value provided crack lengths were limited to between c/d values of 0.34–0.6.

Extrapolation of the measured values of critical stress intensity factors of specimens over the porosity range 7–37% using linear, exponential (equation 40) and a two-term polynomial (equation 39) gave values as shown in Table 12. In all cases values obtained from diametral compression were higher than those from edge opening. However, all values are significantly higher than those determined from measurements on beams and plates.

The reasons for this variation in the results from these different test procedures have been discussed by York *et al.* (1990) specifically for microcrystalline cellulose. The problem lies in the difficulties in the introduction of a two-dimensional sharp crack into a specimen and accurate measurement of its length and velocity on the application of load. Ideally, K_{IC} should be independent of crack length and for those materials which exhibit a flat crack growth resistance curve all methods of measurement should produce equivalent data for K_{ICo}. However, many materials, especially ceramics (Munz, 1983) have been shown to exhibit rising crack resistance curves and hence the method of crack induction and notch geometry become critical. These measurements on specimens with sawn or

Table 12 Critical stress intensity factors of microcrystalline cellulose using radial edge cracked tablets (Roberts and Rowe, 1989)

Method	Equation	K_{ICo} (MPam$^{1/2}$)
Edge opening	Linear	1.91
Diametral compression	Linear	2.11
Edge opening	Exponential	2.24
Diametral compression	Exponential	2.98
Edge opening	Polynomial	2.31
Diametral compression	Polynomial	2.35

machined notch will always produce lower values of K_{IC} than those where the crack is introduced by a controlled flaw. This is the case for the double-torsion method and radially edge cracked tablet as in these methods the total amount of crack extension at maximum load is always higher. This is direct evidence that microcrystalline cellulose has a rising crack growth resistance curve (Roberts and Rowe, 1989) and that this is the reason for the variation.

Vickers indentation cracking test

Interest in indentation fracture goes back to the empirical finding (Auerbach, 1891) that the load required to form a crack is proportional to the radius of the spherical indenter for indentations of radius $\ll 1$ cm. Although spherical and pyramidal indenters are both used, only the latter will be discussed as the former method is applicable only to transparent material (e.g. glasses). The most distinctive feature of the indentation of brittle materials by the Vickers or pyramidal indenter is the appearance of cracks emanating from the corners of the indent (Fig. 18) and it is from the measurement of the lengths of these cracks that it is possible to determine the critical stress intensity factor for indentation cracking or K_c.

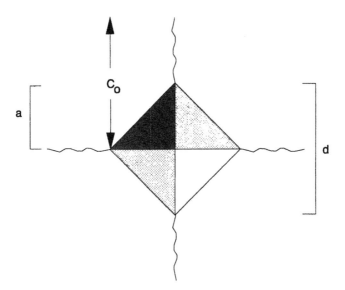

Fig. 18 Schematic diagram showing cracking around a Vickers indent. C_o, crack length; d, length of diagonal of indent; a, length of half-diagonal of indent.

Before describing the technique it is important to realize the assessment of K_c is strictly semiempirical because, first, the analytical solutions of the stress field around indentations have not been solved and only approximate solutions have been derived, and second, the deformation field is not homogeneous and anisotropy and fracture complicate the problem.

The most common solution is an equation based on the model of Evans and Charles (1976) and from experimental observations of Marshall and Lawn (1979):

$$K_c = \chi \frac{F}{C_o^{3/2}} \tag{44}$$

where F is indentation load, C_o is the crack length (as in Fig. 18) and χ is a constant which can include terms reflecting the plastoelasticity of the system, e.g. $(E/H)^m$ where m is a constant which varies depending on the calibrating materials assuming that $K_c = K_{IC}$, where K_{IC} is from conventional fracture testing. Obviously, the wider the range of solids used, the more universal is the equation and in this respect Evans and Charles (1976) have used ceramic data from the double torsion test as calibrating data obtaining a value of χ of 0.0824.

The most widely used equation for determining K_c is from Antis et al. (1981) using both measured and literature data for H, E and K_{IC} (double cantilever beam test) for ceramics to allow calibration:

$$K_c = \Phi \left(\frac{E}{H}\right)^{1/2} \frac{F}{C_o^{3/2}} \tag{45}$$

where Φ is a calibration constant equivalent to 0.016.

In a review article Ponton and Rawlings (1989a,b) analysed 19 indentation fracture equations for their ability to equate to conventional testing and found an equation by Lankford (1982) the most universal:

$$K_c = \Phi \left(\frac{E}{H}\right)^{2/5} \left(\frac{F}{a^{3/2}}\right) \left(\frac{a}{C_o}\right)^{1.56} \tag{46}$$

where Φ is a calibration constant equivalent to 0.0363 and a is the half diagonal or $a = d/2$.

Only two papers have examined the indentation fracture test as a means of determining the critical stress intensity factor of pharmaceuticals (Duncan-Hewitt and Weatherly, 1989a,b). In the first of these, Duncan-Hewitt and Weatherly (1989a) evaluated the indentation test using sucrose crystals. Microindentation was performed using a Leitz–Wetzlar Miniload hardness tester (Vickers pyramidal diamond indenter), applying loads of 147 mN, with the indentations and cracks measured using a light microscope (Leitz). Large crystals were prepared (1–4 mm diameter) by slow evaporation of a saturated aqueous solution at 23°C over a period of 3–6 months. The prismatic crystals were washed in ethanol to remove traces of crystallization solution and were stored under controlled conditions before testing. Furthermore, some specific crystal faces – (100), (010) and (001) – were prepared by either abrading with decreasing grades of emery paper or by cleavage. Crystals were mounted in plasticine prior to testing. The authors found that fractures appeared anisotropic and that if crystals were tested immediately after polishing, fracture was either suppressed or significantly decreased, e.g. crack lengths after 2 min, 4 min and 20 min polishing were $0\,\mu m$, $16\,\mu m$ and $34\,\mu m$, respectively.

In addition to using equation 45 of Antis *et al.* (1981), Duncan-Hewitt and Weatherly (1989a) examined the equation of Laugier (1987):

$$K_c = \Phi \left(\frac{E}{H}\right)^{2/3} \left(\frac{F}{C_o^{3/2}}\right) \left(\frac{a}{l}\right)^{1/2} \qquad (47)$$

where Φ is a calibration constant equivalent to 0.0143 and $l = C_o - a$.

The authors found that the values of K_c using the two equations were similar (the mean for the various faces are given in Table 13). Furthermore, they reported that equation 47 (Laugier, 1987) appeared to emphasize the apparent fracture anisotropy. The fracture plane with the lowest value was the (100) in agreement with the easiest to cleave plane (although the (101) had the lowest K_c value) and the plane with greatest K_c was the (001) plane. It is interesting to note that the (100) plane had the hardest surface whereas the (001) plane had the softest (Table 6).

Calculations have been performed to analyse the ability of all the fracture indentation equations (see Table 13) to equate to the value of sucrose from flexure testing (Roberts *et al.*, 1993), e.g. the equivalence of K_c to K_{ICo},

Table 13 Critical stress intensity factors for sucrose

Method	Critical stress intensity factor (MPam$^{1/2}$)	Reference
Indentation	0.061	Evans and Charles (1976)
	0.078	Antis *et al.* (1981)
	0.089	Laugier (1987)
	0.104	Lankford (1982)
Single edge notched beam	0.224	Roberts *et al.* (1993)

and these data are presented in Table 13. In all cases the indentation technique gives values of the critical stress intensity considerably lower than the flexure method, with the value calculated using the Lankford (1982) equation giving the closest agreement and confirming the study by Ponton and Rawlings (1989a,b) using ceramics.

In a later paper Duncan-Hewitt and Weatherly (1989b) published further indentation critical stress intensity factors calculated from the Antis *et al.* (1981) relationship (equation 45). These are shown in Table 14 with the corresponding data from single edge notched beam (three point) beam testing (Roberts *et al.*, 1993). Although the rank order is the same the differences in magnitude of the values are large (with the exception of sodium chloride). A possible explanation for these differences is that in the indentation test the theory is not exact and the equations are derived from calibration with ceramics, i.e. materials with plastoelastic properties considerably different from pharmaceutical materials.

Table 14 Comparison of critical stress intensity factors measured using indentation (K_c) and single-edge notched beams (K_{ICo}).

Material	K_c (MPam$^{1/2}$)	K_{ICo} (MPam$^{1/2}$)
Sodium chloride	0.50	0.48
Sucrose	0.08	0.22
Paracetamol	0.05	0.12
Adipic acid	0.02	0.14

Despite its shortcomings the indentation technique has certain advantages over other testing methods as it can be used on small samples, e.g. single crystals, specimen preparation is relatively simple and the indentation hardness and Young's modulus can be determined simultaneously.

RELATIONSHIPS BETWEEN PROPERTIES

All the properties described above are interrelated with specific ratios between them associated with second-order characteristics such as brittleness, ductility, plasticity and toughness. In this section a number of relationships are reviewed and their usefulness examined.

Of specific interest is the semiempirical equation derived by Marsh (1964) while studying Vickers indentation:

$$\frac{H}{\sigma_y} = 0.07 + 0.61 \ln \left(\frac{E}{\sigma_y}\right) \tag{48}$$

From this relationship (Fig. 19) and a knowledge of the materials studied it is possible to show: first that if E/σ_y between 10–25 then H/σ_y is between 1.5–2.0, corresponding to highly elastic materials typical of many polymers; second if E/σ_y is between 25–30 then H/σ_y is between 2.0–2.2, corresponding to brittle materials typical of glasses (if $H/\sigma_y > 2.2$ then there is a tendency towards a reduction in brittle behaviour); and third if $E/\sigma_y > 150$ then $H/\sigma_y > 3.0$, corresponding to a rigid–plastic material typical of metals.

This equation is of specific interest as it implies that the ratio H/σ_y is not always equal to 3 as suggested earlier (equation 5) but is related to both the plasticity and elasticity of the material. This would account for the small differences seen in the calculated yield stress from indentation hardness measurements (using a value of 3) and those from Heckel plots shown in Table 7. This ratio has been extensively examined by Roberts and Rowe (1987a) and for the majority of pharmaceutical materials shown to lie between 2.3 and 3.4.

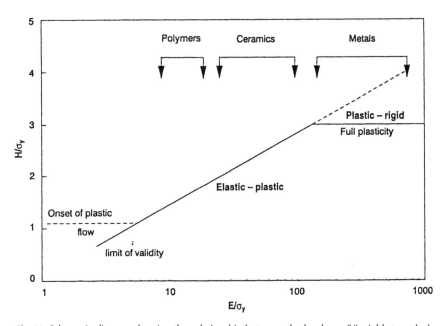

Fig. 19 Schematic diagram showing the relationship between the hardness (H), yield stress (σ_y), and Young's modulus of elasticity (E) for a range of materials (adapted from Tabor, 1979).

Equation 48 can be useful in a predictive capacity because it is possible, from a knowledge of two of the properties, to calculate the third with a reasonable degree of accuracy (Roberts and Rowe, 1987a).

A modification of equation 48 allows for the use of both spherical and pyramidal indenters (Johnson, 1970):

$$\frac{H}{\sigma_y} = \frac{2}{3}\left(1 + \ln\left(\frac{EX}{3\sigma_y}\right)\right) \tag{49}$$

where X is either $\cot\theta$, where θ is the semiapical angle for a pyramid, or d/D, where d and D are the diameters of the indent and sphere, respectively.

Using a similar approach to Marsh (1964) and Johnson (1970), Marshall *et al.* (1982) showed that the ratio H/E is also an important factor in indentation as elastic materials always exhibit a high value whereas plastic materials always exhibit a low value. This ratio has been extrapolated to pharmaceutical materials by Hiestand and Smith (1984) and renamed the 'strain index'; it is described as indicating the relative strain during elastic recovery that follows plastic deformation or the relative potential for strain energy to develop at the tip of a defect. Both these and other workers (Duncan-Hewitt and Weatherly, 1989b) have suggested that the strain index is useful to predict tablettability in terms of capping and lamination but comparison of literature data (Table 15) would tend to dispute this claim as some materials with similar strain indices exhibit different compaction behaviour.

Table 15 The strain index, H/E, for some excipients and drugs

Material	Strain index	Comments	References
Paracetamol	0.05	Lamination/capping	Duncan-Hewitt and Weatherly (1989b)
Erythromycin	0.04	Lamination	Hiestand and Smith (1984)
Adipic acid	0.03	Lamination	Duncan-Hewitt and Weatherly (1989b)
Avicel PH101	0.025	Excellent	Hiestand and Smith (1984)
Starch	0.023	Good	Hiestand and Smith (1984)
Lactose S.D.	0.021	Poor	Hiestand and Smith (1984)
Sucrose	0.02	–	Duncan-Hewitt and Weatherly (1989b)
Sucrose	0.016	Splitting	Hiestand and Smith (1984)
Phenacetin	0.013	Lamination	Hiestand and Smith (1984)
Ibuprofen	0.011	Good	Hiestand and Smith (1984)
Hexamine	0.006	Capping	Hiestand and Smith (1984)
NaCl	0.005	Good	Duncan-Hewitt and Weatherly (1989b)

Table 16 Brittleness indices for some excipients and drugs

Material	H (MPa)	K_{ICo} (MPam$^{1/2}$)	$\dfrac{H}{K_{ICo}}$ (μm$^{-1/2}$)
Avicel PH101	168[a]	0.7569	0.22
Lactose β anhydrous	251[b]	0.7597	0.33
Ibuprofen	35[b]	0.1044	0.34
Sodium chloride	213[c]	0.4769	0.45
Aspirin	87[d]	0.1561	0.56
Adipic acid	123[c]	0.1398	0.88
Paracetamol DC	265[a]	0.2463	1.08
α-Lactose monohydrate	515[b]	0.3540	1.45
Sucrose	645[c]	0.2239	2.88
Paracetamol	421[c]	0.1153	3.65

[a] Jetzer et al. (1983a)
[b] Leuenberger (1982)
[c] Duncan-Hewitt and Weatherly (1989b)
[d] Ridgway et al. (1969a)

By considering that the brittleness of a material is a measure of the relative susceptibility of that material to the two competing mechanical responses of deformation and fracture, Lawn and Marshall (1979) proposed a brittleness index defined as the ratio of indentation hardness to the critical stress intensity factor with units μm$^{-1/2}$. Table 16 shows data for a variety of materials; H is obtained from indentation hardness and K_{ICo} from single edge notched beam tests (Roberts et al., 1993). Duncan-Hewitt and Weatherly (1989b) also used this index suggesting that materials with a high value tend to undergo fragmentation during compaction. The results in Table 16 for a wider range of materials tend to support this approach with materials (e.g. Avicel PH101, ibuprofen) known to be essentially ductile in behaviour exhibiting very low values and materials (e.g. sucrose, paracetamol) known to be brittle in behaviour exhibiting very high values.

It is interesting to compare the values of the brittleness index given in Table 16 with values in the literature for medium-strength steel (0.1), polymethyl methacrylate (0.14), alumina (3.0) and glass (8.8). Based on this classification all pharmaceutical materials can be considered to be brittle or semi-brittle in nature.

A relationship already proposed earlier (equation 6) between the critical stress intensity factor and Young's modulus of elasticity provides a comparative measure of the fracture toughness (R). Data on a wide range of

Table 17 Fracture toughness values for some excipients and drugs

Material	Young's modulus (GPa)	Critical stress intensity factor (MPam$^{1/2}$)	Fracture toughness (J m^{-2})
Celluloses			
Avicel PH105	9.4[a]	1.33[h]	214.5
Avicel PH105	10.1[b]	1.33[h]	175.1
Avicel PH101	7.8[c]	0.76[i]	643.3
Avicel PH101	10.3[d]	1.21[d]	142.1
Avicel PH101	9.2[a]	0.87[h]	106.5
Avicel PH101	9.0[b]	0.87[h]	84.1
Avicel PH101	7.8[c]	2.24[j]	74.1
Avicel PH102	8.7[b]	0.76[h]	95.2
Avicel PH102	8.2[a]	0.76[h]	70.4
Emcocel	7.1[a]	0.80[h]	90.1
Emcocel	9.0[b]	0.80[h]	94.0
Emcocel (90M)	8.9[a]	0.83[h]	77.4
Emcocel (90M)	9.4[b]	0.83[h]	68.1
Unimac (MG100)	8.0[a]	1.05[h]	137.8
Unimac (MG100)	8.8[b]	1.05[h]	72.7
Unimac (MG200)	7.3[a]	1.05[h]	79.1
Unimac (MG200)	8.0[b]	1.05[h]	56.1
Sugars			
α-Lactose monohydrate	24.1[e]	0.35[i]	5.1
Sorbitol instant	45.0[f]	0.47[f]	4.9
Sucrose	32.3[g]	0.08[g]	0.2
Drugs			
Ibuprofen	5.0[e]	0.10[i]	2.0
Aspirin	7.5[e]	0.16[i]	3.4
Paracetamol DC	11.7[e]	0.25[i]	5.3
Paracetamol	11.7[e]	0.12[i]	1.2
Paracetamol	8.4[g]	0.05[g]	0.3
Others			
Sodium chloride	43.0[g]	0.50[g]	5.8
Adipic acid	4.1[g]	0.02[g]	0.1

[a] Bassam *et al.* (1990)
[b] Bassam *et al.* (1988)
[c] Roberts *et al.* (1989b)
[d] Mashadi and Newton (1987b)
[e] Roberts *et al.* (1991)
[f] Mashadi and Newton (1987a)
[g] Duncan-Hewitt and Weatherly (1989b)
[h] York *et al.* (1990)
[i] Roberts *et al.* (1993)
[j] Roberts and Rowe (1989)

materials using beam specimens to generate values for modulus of elasticity and critical stress intensity factors are shown in Table 17. It can be seen that fracture toughness can vary over two orders of magnitude, with materials exhibiting high values being regarded as tough.

Lawn and Wilshaw (1975) and Atkins and Mai (1985) have used fracture toughness as a means of classifying materials, e.g.

- For highly brittle materials where crack propagation occurs via the reversible fracture of cohesive bonds the fracture energy is equal to the surface free energy $R = 0.5$–$5 \, \mathrm{J \, m}^{-2}$.
- For semi-brittle materials where there is some plastic flow at crack tips $R = 5$–$50 \, \mathrm{J \, m}^{-2}$.
- For non-brittle materials where blunting of crack tips occur R can be in excess of $5 \times 10^4 \, \mathrm{J \, m}^{-2}$.

EFFECT OF PARTICLE SIZE – BRITTLE–DUCTILE TRANSITIONS

It has been known for many years that if Heckel plots were constructed for a material of varying particle size the reciprocal of the gradient over the central linear portion of the graph (now defined as the deformation stress, σ_d) varied either remaining constant or increasing as particle size decreased (Hersey and Rees, 1970; York, 1978). An extensive study (Roberts and Rowe, 1986, 1987b; Roberts et al., 1989a) showed that the effect of particle size was even more complex, depending on the material under test (Fig. 20). For a material known to undergo plastic deformation (e.g. microcrystalline cellulose) no effect of particle size could be seen; for a material known to undergo brittle fracture (e.g. dolomite), the deformation stress increased with decreasing particle size and for materials known to undergo a combination of brittle fracture and plastic deformation (e.g. α-lactose monohydrate) the deformation stress increased with decreasing particle size to a plateau value.

A comparison of the shapes of the curves with that predicted schematically in Fig. 1 shows that the point of change is indicative of a brittle–ductile transition equivalent to the critical particle size d_{crit} predicted from theory. This has been confirmed for sodium chloride using independent measurements of yield stress and critical stress intensity factor (Roberts et al., 1989a).

Calculated critical sizes (equation 4) for a variety of excipients and drugs using yield stress data and critical stress intensity factors given in previous tables are shown in Table 18. It can be seen that critical particle sizes can vary over several orders of magnitude. The values appear reasonable in the

Plate 1 Punch and die set used to prepare beams.

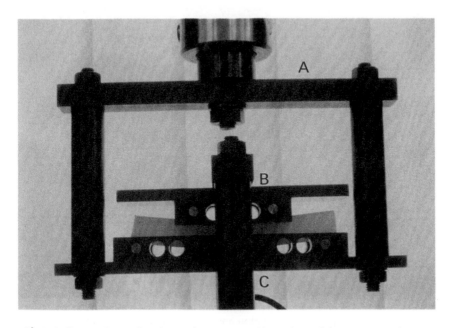

Plate 2 Four-point testing rig used to measure Young's modulus. A, upper frame; B, block; C, lower frame (see text).

Plate 3 Three-point testing rig used to measure Young's modulus.

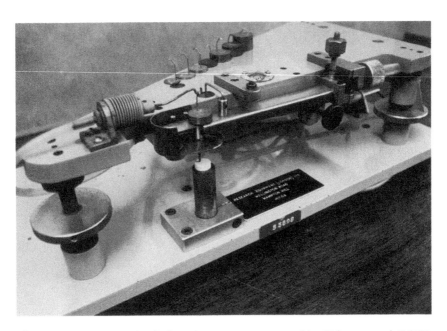

Plate 4 A pneumatic microindentation apparatus as used by Ridgway *et al.* (1970).

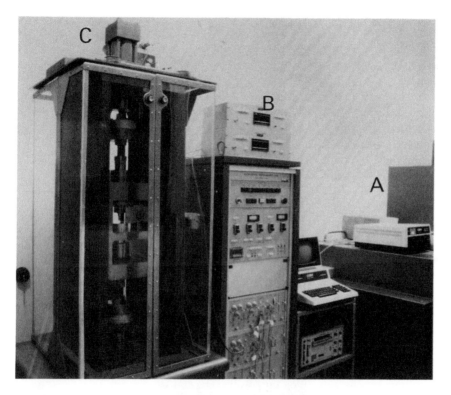

Plate 5 Tablet compression simulator as used by Roberts and Rowe (1985); A, hydraulic power pack; B, electronic control unit; C, loading frame.

Plate 6 Three point testing rig used to measure critical stress intensity factor.

Plate 7 Precracking of a tablet using a scalpel blade.

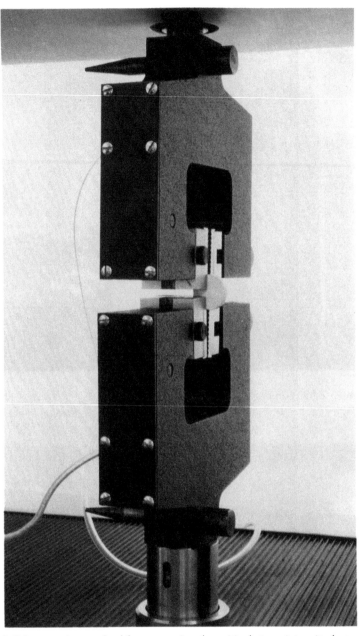

Plate 8 Edge opening method for measuring the critical stress intensity factor for a radially edge-cracked tablet.

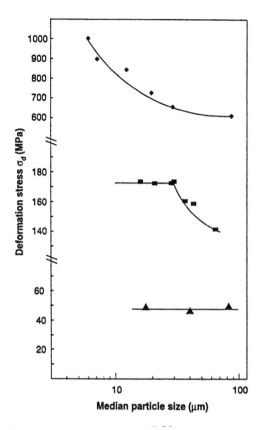

Fig. 20 The effect of particle size on the deformation stress as measured using Heckel plots for: ♦, dolomite; ■, α lactose monohydrate; ▲, microcrystalline cellulose (adapted from Roberts and Rowe, 1987b).

light of experimental findings for microcrystalline cellulose and α-lactose monohydrate in Fig. 20. The value for microcrystalline cellulose is similar to that determined for other polymeric materials (Kendall, 1978). The differences in the two values for the two lactoses are consistent with the findings of workers describing their compaction properties (Vromans *et al.*, 1986). Of the three drugs paracetamol has the lowest critical particle size, which is consistent with it being very brittle. However, the addition of a polymeric binder to paracetamol (i.e. paracetamol DC) increases its critical particle size by an order of magnitude causing it to become plastic in nature.

Table 18 Critical particle size for some excipients and drugs

Material	H (MPa)	K_{ICo} (MPam$^{1/2}$)	d_{crit} (μm)
Microcrystalline cellulose	168[a]	0.7569	1949
Lactose β anhydrous	251[b]	0.7597	873
Ibuprofen	35[b]	0.1044	854
Aspirin	87[c]	0.1561	309
Paracetamol DC	265[a]	0.2463	83
α-Lactose monohydrate	515[b]	0.3540	45
Sucrose	645[d]	0.2239	12
Paracetamol	421[d]	0.1153	7

[a] Jetzer et al. (1983a)
[b] Leuenberger (1982)
[c] Ridgway et al. (1969a)
[d] Duncan-Hewitt and Weatherly (1989b)

EFFECT OF MOISTURE CONTENT

The moisture content of a material can affect its mechanical properties generally by acting as an internal 'lubricant' facilitating slippage and plastic flow (Khan et al., 1981). Obviously, this will lead to a decrease in both yield stress and Young's modulus of elasticity as adequately demonstrated for microcrystalline cellulose (Roberts and Rowe, 1987d; Bassam et al., 1990). In the case of yield stress, increasing the moisture content caused a linear decrease independent of the source and batch of the material under test (Fig. 21). Similar results have been recorded for paracetamol (Garr and Rubinstein, 1992).

The effect of moisture content on the critical stress intensity factor is more complex. Bassam et al. (1990) have shown that for microcrystalline cellulose there is a decrease with increasing moisture content, i.e. the material became less resistant to crack propagation. Similar results have been found for glass by Wiederhorn (1967) who suggested that the effect was caused by water interacting at the crack tip.

PREDICTION OF COMMINUTION BEHAVIOUR

It is evident from the previous discussion that the critical particle size, d_{crit}, is an important factor in predicting comminution behaviour as, for a material known to undergo brittle fracture, it should be the limiting size obtained on hammer milling. Unfortunately, data on limiting sizes from

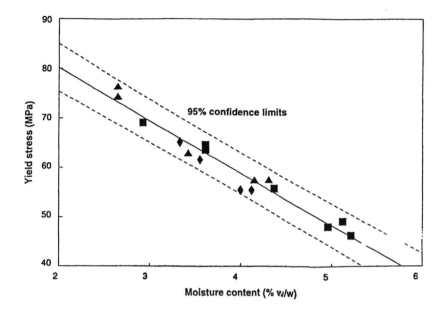

Fig. 21 The effect of moisture content on the yield stress of microcrystalline cellulose for: ■, Avicel; ▲, Unimac; ♦, Emcocel (adapted from Roberts and Rowe, 1987d).

hammer milling are few. However, Snow *et al.* (1988) have reported values for sugar (sucrose) and calcium carbonate of 19 μm and 5 μm, respectively. These values are close to the d_{crit} values calculated from equation 4 of 12 μm and 1 μm (Kendall, 1978), respectively.

PREDICTION OF COMPACTION BEHAVIOUR

At present a fundamental knowledge and theory describing the compaction of particulate solids, specifically pharmaceuticals, is lacking. While it is generally recognized that the mechanical properties of such materials are critical in directing compaction behaviour, there is no predictive capability. Roberts and Rowe (1987a) presented a pragmatic approach to predicting the consolidation mechanism of pharmaceutical materials based on a knowledge of Young's modulus of elasticity, yield stress, hardness and strain rate sensitivity (Fig. 22). This approach, combined with further measurements on critical stress intensity factor enabling the prediction of critical particle sizes, d_{crit} (to account for particle size effects), will enable prediction of

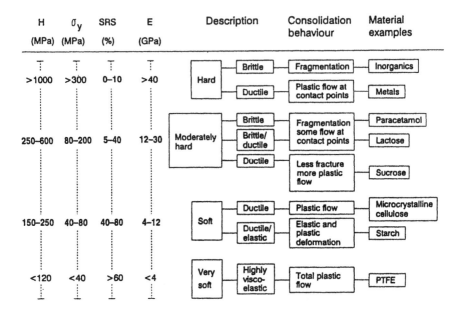

Fig. 22 The relationship between material properties, hardness (H), yield stress (σ_y), strain rate sensitivity (SRS) and Young's modulus of elasticity (E), and compaction behaviour (adapted from Roberts and Rowe, 1987a).

the consolidation mechanism of any material of known particle size. This knowledge will allow tablet formulators to take a more scientific approach to formulation. Indeed, a combination of a predictive approach based on mechanical property measurements combined with heuristics should enable the development of quality expert systems for tablet formulation.

REFERENCES

Al-Hassani, S. T. S., Tennant, P. W. and Sarumi, M. A. (1989). *Powder Metall.* **32**, 204–208.

Antis, G. R., Chantikul, P., Lawn, B. R. and Marshall, D. B. (1981). *J. Am. Ceram. Soc.* **64**, 533–538.

Atkins, A. G. and Mai, Y. W. (1985). *Elastic and Plastic Fracture, Metals, Polymers, Ceramics, Composites, Biological Materials*, Ellis Horwood-Halsted-Wiley, Chichester.

Auerbach, F. (1891). *Ann. Phys. Chem.* **43**, 61–69.

Aulton, M. E. (1977). *Manufact. Chem. Aerosol News*, **48**, 28–36.

Aulton, M. E. (1981). *Pharm. Acta Helv.* **56**, 133–136.

Aulton, M. E. and Marok, I. S. (1981). *Int. J. Pharm. Tech. Prod. Mfr.* **2**, 1–6.

Aulton, M. E. and Tebby, H. G. (1975). *J. Pharm. Pharmacol.* **27**, Suppl., 4P.

Aulton, M. E., Tebby, H. G. and White, P. J. P. (1974). *J. Pharm. Pharmacol.* **26**, Suppl. 59P–60P.

Bassam, F., York, P., Rowe, R. C. and Roberts, R. J. (1988). *J. Pharm. Pharmacol.* **40**, Suppl., 68P.

Bassam, F., York, P., Rowe, R. C. and Roberts, R. J. (1990). *Int. J. Pharm.* **64**, 55–60.

Bassam, F., York, P., Rowe, R. C. and Roberts, R.J. (1991). *Powder Tech.* **65**, 103–111.

Bateman, S. D., Rubinstein, M. H. and Wright, P. (1987). *J. Pharm. Pharmacol.* **39**, Suppl. 66P.

Bateman, S. D., Rubinstein, M.H., Rowe, R. C., Roberts, R. J., Drew, P. and Ho, A. Y. K. (1989). *Int. J. Pharm.* **49**, 209–212.

Breval, E. and Macmillan, N. H. (1985). *J. Mater. Sci. Lett.* **4**, 741–742.

Brinell, J. A. (1901). *J. Iron & Steel Inst.* **59**, 243–251.

British Standards Institution (1977). *Plane strain fracture toughness* (K_{IC}) *of metallic materials, BS 5447*. BSI publications, London.

Brown, W. F. and Srawley, J. E. (1966). *Plane Strain Crack Toughness Testing of High Strength Metallic Materials*. ASTM Special Tech. Publ. No. 410.

Church, M. S. (1984). Mechanical characterisation of pharmaceutical powder compacts. Ph.D. thesis, Nottingham.

Cottrell, A. H. (1964). *The Mechanical Properties of Matter*. John Wiley and Sons, London.

Danjo, K., Ertell, C. and Cartensen, J. T. (1989). *Drug Dev. Ind. Pharm.* **15**, 1–10.

Dean, E. A. and Lafez, J. A. (1983). *J. Amer. Ceram. Soc.* **66**, 366–370.

DeBoer, H., Bolhuis, G. K. and Lerk, C. F. (1978). *Powder Technol.* **20**, 75–82.

Duberg, M. and Nystrom, C. (1986). *Powder Technol.* **46**, 67–75.

Duncan-Hewitt, W. C. and Weatherly, G. C. (1989a). *Pharm. Res.* **6**, 373–378.

Duncan-Hewitt, W. C. and Weatherly, G. C. (1989b). *J. Mat. Sci. Lett.* **8**, 1350–1352.

Ejiofor, O., Esezobo, S. and Pilpel, N. (1986). *J. Pharm. Pharmacol.* **38**, 1–7.

Evans, A. G. (1974). In *Fracture Mechanics of Ceramics, Volume 1 Concepts, Flaws and Fractography* (Bradt, R. C., Hasselman, D. P. H. and Lange, F. F., eds) pp. 17–48. Plenum Press, New York.

Evans, A. G. and Charles, E. A. (1976). *J. Am. Ceram. Soc.* **59**, 371–372.

Fell, J. T. and Newton, J. M. (1971). *J. Pharm. Sci.* **60**, 1866–1869.

Galli, B. and Leuenberger, H. (1986). VDI Ber., **583**, 173–197.

Garr, J. S. M. and Rubinstein, M. H. (1992). *Int. J. Pharm.* **81**, 187–192.

Heckel, R. W. (1961a). *Trans. Metall. Soc. A.I.M.E.* **221**, 671–675.

Heckel, R. W. (1961b). *Trans. Metall. Soc. A.I.M.E.* **221**, 1001–1008.

Hersey, J. A. and Rees, J. E. (1970). *Particle Size Analysis*, pp. 33–41. Society for Analytical Chemistry, Bradford.

Hiestand, E. N. (1985). *J. Pharm. Sci.* **74**, 768–770.

Hiestand, E. N. and Smith, D. P. (1984). *Powder Technol.* **38**, 145–159.

Hiestand, E. N., Amidon, G. E., Smith, D. P. and Tiffany, B. D. (1981). *Proc. Tech. Program: Int. Powder Bulk Solids Handl. Process.* 383–387.

Hiestand, E. N., Bane, J. M. Jnr. and Strzelinski., E. P. (1971). *J. Pharm. Sci.* **60**, 758.

Humbert-Droz, P., Gurny, R., Mordier, D. and Doelker, E. (1983). *Int. J. Pharm. Tech. & Prod. Mfr.* **4**, 29–35.

Humbert-Droz, P., Mordier, D. and Doelker, E. (1982). *Pharm. Acta Helv.* **57**, 136–143.

Hussain, M. S. H., York, P. and Timmins, P. (1991). *Int. J. Pharm.* **70**, 103–109.

Ichikawa, J., Imagawa, K. and Kaneniwa, N. (1988). *Chem. Pharm. Bull.* **36**, 2699–2702.

Jetzer, W., Leuenberger, H. and Sucker, H. (1983a). *Pharm. Tech.* **7** (4), 33–39.

Jetzer, W., Leuenberger, H. and Sucker, H. (1983b). *Pharm. Tech.* **7** (11), 33–48.

Jetzer, W. E., Johnson, W. B. and Hiestand, E. N. (1985). *Int. J. Pharm.* **26**, 329–337.

Johnson, K. L. (1970). *J. Mech. Phys. Solids*, **18**, 115–126.

Kendall, K. (1978). *Nature* **272**, 710–712.

Kendall, K. and Gregory, R. D. (1987). *J. Mater. Sci.*, **22**, 4514–4517.

Kendall, K., Alford, M. McN. and Birchall, J. D. (1987). *Proc. R. Soc. Lond.* **A412**, 269–283.

Kerridge, J. C. and Newton, J. M. (1986). *J. Pharm. Pharmacol.* **38**, Suppl., 78P.

Khan, K. A., Musikabhumma, P. and Warr, J.P. (1981). *Drug Dev. Ind. Pharm.* **7**, 525–538.

Kies, J. A. and Clark, A. B. J. (1969). In *Proc. 2nd Int. Conf. Fracture, Brighton*, (Pratt, P. L., ed.) pp. 483–491. The Powder Advisory Centre, London.

Kim, Y., Machida, K., Taga, T. and Osaki, K. (1985). *Chem. Pharm. Bull.* **33**, 2641–2647.

Lankford, J. (1982). *J. Mater. Sci. Lett.* **1**, 493–495.

Laugier, M. T. (1987). *J. Mater. Sci. Lett.* **6**, 355–356.

Lawn, B. R. and Marshall, D. B. (1979). *J. Am. Ceram. Soc.* **62**, 347–350.

Lawn, B. R. and Wilshaw, T. R. (1975). *Fracture of Brittle Solids*, pp. 109–112. Cambridge University Press, Cambridge.

Leuenberger, H. (1982). *Int. J. Pharm.* **12**, 41–55.

Leuenberger, H. and Rohera, B. D. (1985). *Pharm. Acta Helv.* **60**, 279–286.

Leuenberger, H. and Rohera, B. D. (1986). *Pharm. Res.* **3**, 12–22.

Marsh, D. M. (1964). *Proc. Roy. Soc.* **A279**, 420–435.

Marshall, D. B. and Lawn, B. R. (1979). *J. Mater. Sci.* **14**, 2001–2012.

Marshall, D. B., Noma, T. and Evans, A. G. (1982). *Commun. Am. Ceram. Soc.* **65**, C175–C176.

Mashadi, A. B. and Newton, J. M. (1987a). *J. Pharm. Pharmacol.* **39**, Suppl., 67P.

Mashadi, A. B. and Newton, J. M. (1987b). *J. Pharm. Pharmacol.* **39**, 961–965.

Mashadi, A. B. and Newton, J. M. (1988). *J. Pharm. Pharmacol.* **40**, 597–600.

Munz, D. (1983). In *Fracture Mechanics of Ceramics, Volume 6 Measurements, Transformations, and High-temperature Fracture*, (Bradt, R. C., Evans, A. G., Hasselman, D. P. H. and Lange, F. F., eds) pp. 1–26. Plenum Press, New York.

Nutter Smith, A. (1949). *Pharm. J.* **163**, 477–478.

Outwater, J. O. and Jerry, D. J. (1966). On the fracture energy of glass, N.R.L. Interim Genet Report, contract No. NR3219 (01) (X), University of Vermont, Burlington, Vermont, AD 64084.

Paronen, P. (1987). In *Pharmaceutical Technology: Tabletting Technology*, (Rubinstein, M. H., ed.) pp. 139–144. Ellis Horwood Ltd., Chichester.

Podczeck, von F. and Wenzel, U. (1989). *Pharm. Ind.* **51**, 524–527.

Ponton, C. B. and Rawlings, R. D. (1989a). *Mater. Sci. Technol.* **5**, 865–872.

Ponton, C. B. and Rawlings, R. D. (1989b). *Mater. Sci. Technol.* **5**, 961–976.

Puttick, K. E. (1980). *J. Phys. D: Appl. Phys.* **12**, 2249–2262.

Radebaugh, G. W., Babu, S. R. and Bondi, J. N. (1989). *Int. J. Pharm.* **57**, 95–105.

Ragnarsson, G. and Sjogren, J. (1984). *Acta. Pharm. Suec.* **21**, 141–144.

Ragnarsson, G. and Sjogren, J. (1985). *J. Pharm. Pharmacol.* **37**, 145–147.

Ramberger, R. and Burger, A. (1985). *Powder Technology*, **43**, 1–9.

Rees, J. E. and Rue, P. J. (1978). *J. Pharm. Pharmacol.* **30**, 601–607.

Ridgway, K., Aulton, M. E. and Rosser, P. H. (1970). *J. Pharm. Pharmacol.* **22**, 70S–78S.

Ridgway, K., Shotten, E. and Glasby, J. (1969a). *J. Pharm. Pharmacol.* **21**, Suppl. 19S–23S.

Ridgway, K., Glasby, J. and Rosser, P.H. (1969b). *J. Pharm. Pharmacol.* **21**, Suppl. 24S–29S.

Roberts, R. J. (1991). The elasticity, ductility and fracture toughness of pharmaceutical powders. Ph.D thesis, Bradford.

Roberts, R. J. and Rowe, R. C. (1985). *J. Pharm. Pharmacol.* **37**, 377–384.

Roberts, R. J. and Rowe, R. C. (1986). *J. Pharm. Pharmacol.* **38**, 567–571.

Roberts, R. J. and Rowe, R. C. (1987a). *Chem. Eng. Sci.* **42**, 903–911.

Roberts, R. J. and Rowe, R. C. (1987b). *Int. J. Pharm.* **36**, 205–209.

Roberts, R. J. and Rowe, R. C. (1987c). *Int. J. Pharm.* **37**, 15–18.

Roberts, R. J. and Rowe, R. C. (1987d). *J. Pharm. Pharmacol.* **39**, Suppl. 70P.

Roberts, R. J. and Rowe, R. C. (1989). *Int. J. Pharm.* **52**, 213–219.

Roberts, R. J., Rowe, R. C. and Kendall, K. (1989a). *Chem. Eng. Sci.* **44**, 1647–1651.

Roberts, R. J., Rowe, R. C. and York, P. (1989b). *J. Pharm. Pharmacol.* **41** Suppl., 30P.

Roberts, R. J., Rowe, R. C. and York, P. (1991). *Powder Technology* **65**, 139–146.

Roberts, R. J., Rowe, R. C. and York, P. (1993). *Int. J. Pharm.* **91**, 173–182.

Sarumi, M. A. and Al-Hassani, S. T. S. (1991). *Powder Technology* **65**, 51–59.

Simmons, G. and Wang, H. (1971). *Single Crystal Elastic Constants and Calculated Aggregate Properties: A Handbook*, M.I.T., Cambridge.

Smith, R. and Sandland, G. (1925). *Iron & Steel Inst.* **1**, 285–304.

Snow, R. H., Kaye, B. H., Capes, C. E. and Sresty, G. C. (1988). In *Perry's Chemical Engineer's Handbook, Sixth edn*, (Perry, R. H., Green, D. W. and Maloney, J. O., eds) pp. 8.1–8.42. McGraw-Hill Inc, New York.

Spengler, H. and Kaelin, A. (1945). *Pharm. Acta Helv.* **20**, 239–243.

Spinner, S., Knudsen, F. P. and Stone, L. (1963). *J. Res. Nat. Bur. Stand. (Eng. Instr.)* **67C**, 39–46.

Spriggs, J. M. (1961). *J. Am. Ceram. Soc.* **44**, 628–629.

Tabor, D. (1979). In *Physics of Materials*, (Borland, D. W., Clarebrough, L. M. and Moore, A. J. W., eds) pp. 271-282. Univ. Melbourne, Dept. Min. Metall., Parkville.

Tobolsky, A. V. (1962). *Properties and Structure of Polymers*, pp. 1-12. John Wiley and Sons, New York.

Vromans, H. and Lerk, C. F. (1988). *Int. J. Pharm.* **46**, 183-192.

Vromans, H., De Boer, A. H., Bolhuis, G. K. and Lerk, C. F. (1986). *Drug Dev. Ind. Pharm.* **12**, 1715-1730.

Wang, J. C. (1984). *J. Mater. Sci.* **19**, 801-808.

Wiederhorn, S. M. (1967). *J. Am. Ceram. Soc.* **50**, 407-414.

Wong, D. Y. T. and Aulton, M. E. (1987). *J. Pharm. Pharmacol.* **39**, Suppl. 124P.

York, P. (1978). *J. Pharm. Pharmacol.* **30**, 6-10.

York, P. (1979). *J. Pharm. Pharmacol.* **31**, 244-246.

York, P., Bassam, F., Rowe, R. C. and Roberts, R. J. (1990). *Int. J. Pharm.* **66**, 143-148.

2

KINETIC ASPECTS IN THE DESIGN OF PROLONGED ACTION OCULAR DRUG DELIVERY SYSTEMS

A. Urtti

Department of Pharmaceutical Technology, University of Kuopio, P.O. Box 1627, 70211 Kuopio, Finland

INTRODUCTION

Drug treatment of ocular diseases is usually practised as topical ocular drug treatment with eyedrops. Ocular administration is popular because the eye is an easily accessible organ for drug administration. In addition, compared with systemic treatment of ophthalmic diseases (e.g. treatment of glaucoma with oral acetazolamide) smaller doses can be used in local ocular drug administration. Systemic drug treatment of ocular diseases is difficult because the eye is separated from the general blood circulation by two barriers: the blood–aqueous barrier and the blood–vitreous barrier. The properties of these barriers have been reviewed in detail elsewhere (Maurice and Mishima, 1984).

After administration of an eyedrop or ointment the drug is absorbed partly into the eye thereby exerting ocular pharmacological effects such as mydriasis, miosis, decrease of the intraocular pressure, antimicrobial or anti-inflammatory action. Ocular bioavailability after eyedrop administration is poor necessitating higher drug doses which may increase the risk of ocular and systemic side-effects (Lee and Robinson, 1986). Because of the poor bioavailability and rapid drug elimination from the eye, duration of drug action is often short and, for example, pilocarpine eyedrops are administered 3–6 times daily (Havener, 1983; USP DI, 1987). Some kinetic problems of topical ocular drug administration can be solved with prolonged action medications (Shell, 1984).

In some cases drugs are administered to the eye as subconjunctival or

ADVANCES IN PHARMACEUTICAL SCIENCES
ISBN 0–12–032307–9

intravitreous injections (Maurice and Mishima, 1984). These modes of drug administration result in higher intraocular drug concentrations, but they cannot be practised outside the clinics. Also, subconjunctival injections may have low ocular bioavailability because most of the injected drug is absorbed by the systemic circulation (Conrad and Robinson, 1980). This has led to the research of prolonged action subconjunctival dosage forms. Likewise, drug elimination from the vitreous humour can be rapid and there is a need to prolong the action of intravitreally injected drugs. Another goal is to diminish peak drug concentrations and related retinal toxicity after intravitreal injection (Niesman, 1992).

Clearly, prolonged action medications have potential benefits in topical, subconjunctival, and intravitreal drug administration. During the last 20 years many prolonged action drug delivery systems have been tested in the eye. Most drug delivery systems are intended for topical ocular application. They include polymer matrices (Maichuk, 1975; Loucas and Haddad, 1972; Saettone et al., 1984; Urtti et al., 1984; Tang-Liu and Sandri, 1989), reservoir devices (Armaly and Rao, 1973; Sendelbeck et al., 1975), mucoadhesive particles (Hui and Robinson, 1985), liposomes (Lee et al., 1985; Niesman, 1992), nanoparticles (Marchel-Heussler et al., 1990), collagen shields (Friedberg et al., 1991), thermosetting gels (Gurny et al., 1987; Ibrahim et al., 1991), and chemical delivery systems (Lee and Li, 1989). Both polydisperse systems and polymeric inserts have also been tested for subconjunctival and intravitreal administration (Niesman, 1992). Although encouraging results have been published in several studies, very few prolonged action delivery systems are in clinical use.

The performance of the ocular drug delivery system is determined by the in vivo drug release rate from the delivery system, the pharmacokinetics of the released drug and the interactions of the biological system and the delivery system. For example, retention of the delivery system at the site of application and dissolution or erosion of this system in the biological environment are important factors. The ocular physiology/biochemistry and pharmacokinetic properties of each drug set the limits and requirements in the system design. This paper reviews the relevant kinetic factors that should be taken into account when prolonged action dosage forms are designed for topical ocular, subconjunctival or intravitreal use.

PRINCIPLES OF OCULAR PHARMACOKINETICS

Topical ocular drug administration

A schematic and simplified presentation of ocular pharmacokinetics related to topical ocular administration is shown in Fig. 1.

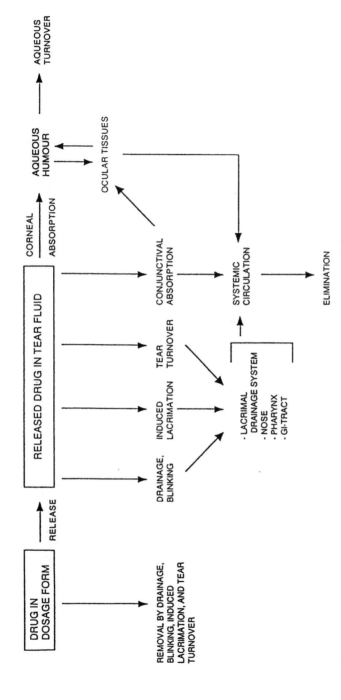

Fig. 1 Schematic presentation of ocular pharmacokinetics related to topical ocular drug administration.

Lacrimal fluid

When eyedrops are administered to the ocular surface the solution mixes with the tear fluid. After administration the extra solution volume rapidly flows from the eye (Chrai *et al.*, 1973, 1974) through the puncta to the lacrimal drainage system and then to the nose, pharynx and gastrointestinal tract (Hurwitz *et al.*, 1975; Chavis *et al.*, 1978). The drainage of the instilled solution is rapid: the drainage rate constant increases with instilled volume – 0.31 min^{-1} and 0.82 min^{-1} for eyedrops of 5 μl and 50 μl, respectively, in rabbits (Chrai *et al.*, 1973). Furthermore, the rate of solution drainage decreases with elevated solution viscosity (Chrai and Robinson, 1974) and mucoadhesiveness (Gurny *et al.*, 1987). Drainage rate of ophthalmic solutions is about three times faster in humans than in rabbits (Zaki *et al.*, 1986). This is partly explained by the higher blinking frequency in humans than in rabbits (Chavis *et al.*, 1978; Saettone *et al.*, 1982).

Another important route of drug loss from the lacrimal fluid is drug absorption to the conjunctiva (Lee and Robinson, 1979; Thombre and Himmelstein, 1984). The conjunctiva of the eye is lined by stratified columnar epithelium two to seven cell layers thick, depending on the conjunctival region (Junqueira *et al.*, 1986). Conjunctival permeability is fairly high so that most drugs diffuse across the conjunctiva easily (Wang *et al.*, 1991). Typical values for conjunctival drug permeabilities are in the order of 1-5 × 10^{-5} cm s^{-1} (Ahmed *et al.*, 1987; Wang *et al.*, 1991). Consequently, nearly 50% of topically instilled pilocarpine is absorbed within a few minutes through conjunctiva to systemic circulation in rabbits (Urtti *et al.*, 1985).

Induced lacrimation may increase the rate of solution drainage from the preocular area for several reasons. Often eyedrop solutions are buffered to a low pH for reasons of stability. Examples of this are pilocarpine, atropine and epinephrine (Dolder and Skinner, 1983). Acidic pH may, however, cause discomfort and thereby increase lacrimation and rate of drug removal from the corneal surface (Conrad *et al.*, 1978). Also, if the osmotic pressure of instilled eyedrops deviates from isotonicity, the rate of instilled solution drainage is increased (Conrad *et al.*, 1978). Topical anaesthesia blocks the induction of lacrimation for the most part and results in a reduced precorneal drainage rate of the instilled solution and in increased ocular pilocarpine absorption in rabbits (Patton and Robinson, 1975). The basal tear turnover has only a minor role in the drug removal from the preocular area (Lee and Robinson, 1979).

At the time of eyedrop instillation there is a concentration gradient between the tear fluid and neighbouring corneal and conjunctival tissues. Consequently, the difference in thermodynamic activity drives drug to these tissues. Thus, for example, in the case of pilocarpine most of the corneal absorption takes place within 3 min after instillation of an eyedrop in rabbits (Chrai *et al.*, 1974). The period of drug absorption is short because the activity gradient decreases rapidly owing to precorneal solution drainage and conjunctival systemic absorption (Chrai *et al.*, 1973; Lee and Robinson, 1979). The initial rate of decrease in lacrimal drug concentration (e.g. pilocarpine) is typically $0.4–1.0 \, min^{-1}$ (Chrai *et al.*, 1973; Lee and Robinson, 1979; Urtti and Salminen, 1985). Thereafter, the drug concentration in the tear fluid decreases much more slowly because pseudo-equilibrium between tear fluid and surrounding tissues has been established (Urtti *et al.*, 1990). In the terminal phase the elimination rate of timolol from the lacrimal fluid of rabbits is even slower $(0.003 \, min^{-1})$ than the rate of normal tear turnover $(0.07 \, min^{-1})$ suggesting considerable back-diffusion from the cornea and conjunctiva to the tear fluid (Urtti *et al.*, 1990). It should be remembered that despite the sustained drug concentrations in the lacrimal fluid after eyedrop administration no more drug absorption takes place during this kinetic phase.

Cornea

The cornea is the main route of drug absorption from the tear fluid to the inner eye in most cases (Doane *et al.*, 1978). Examples of drugs absorbed through this route are timolol (Ahmed and Patton, 1985), pilocarpine (Doane *et al.*, 1978) and hydrocortisone (Doane *et al.*, 1978). For most drugs the main corneal penetration barrier lies in the epithelium (Sieg and Robinson, 1976). The importance of different epithelial cell layers as barriers is dependent on the lipophilicity of compounds (Shih and Lee, 1989). In the case of hydrophilic atenolol the entire epithelium behaves as the penetration-limiting barrier, whereas for timolol and levobunolol, compounds with intermediate lipophilicity, only the most superficial cell layers considerably limit drug permeability (Shih and Lee, 1990). These outermost cellular layers of corneal epithelium are interconnected by tight junctions which limit especially the penetration of hydrophilic molecules (Grass and Robinson, 1988; Wang *et al.*, 1991). In the epithelium wing cells and basal cells are less tightly interconnected and allow intercellular penetration of macromolecules such as horse radish peroxidase (Tonjum, 1974) and fluorescein isothiocyanate (FITC)-labelled poly(L-lysine) (Rojanasakul *et al.*, 1990).

Corneal stroma is a loosely arranged hydrophilic layer with no continuous cellular structure. Consequently, drug diffusion in the stroma is rapid and values are in the typical range not dependent on lipophilicity and molecular weight (Maurice and Mishima, 1984).

Many ophthalmic drugs show considerable lipophilicity that makes trans-
cellular drug penetration possible. Because of the differences in transcel-
lular permeabilities, increasing lipophilicity has been shown to enhance
corneal permeability so that maximal permeability of β-blockers was
observed at log (octanol/water) partition coefficients of 2–3 (Huang *et al.*,
1983). Similar findings were observed with steroids (Schoenwald and Ward,
1978). Data sets of more heterogeneous groups of compounds revealed
substantial deviations from the simple partitioning effects caused by different molecular sizes (Grass and Robinson, 1988) and charges (Liaw *et al.*,
1991). As expected, increased molecular weight decreases the corneal permeability (Grass and Robinson, 1988; Maurice and Mishima, 1984). At
neutral pH the rabbit cornea behaves as if it were negatively charged. Consequently, positively charged L-lysine had a 10 times higher permeability
than the negatively charged L-glutamic acid (Liaw *et al.*, 1991).

In the range of very lipophilic compounds (logP > 3) corneal permeability cannot be further improved by increasing the lipophilicity; steady
(Wang *et al.*, 1991) or decreasing (Schoenwald and Ward, 1978; Huang
et al., 1983) permeabilities at higher lipophilicities have been observed. This
phenomenon is explained by impaired drug desorption from the lipophilic
parts of the corneal epithelium to the hydrophilic stroma (Huang *et al.*,
1983). Therefore, the corneal stroma appears to be the penetration rate
limiting barrier in these cases. It should be remembered, however, that
ocular bioavailability is not directly related to corneal permeability. In addition to the corneal permeability, the ocular bioavailability is affected by the
precorneal drainage factors and by the ratio of permeabilities between the
cornea and conjunctiva (Fig. 1) (Wang *et al.*, 1991).

Corneal permeability of drugs can be improved by optimization of the
formulation pH so that the fraction of the unionized drug is increased
(Francouer *et al.*, 1983; Ashton *et al.*, 1991). Another more demanding
approach is to prepare a prodrug derivative that has improved corneal
absorption characteristics and releases the active parent compound through
enzymatic and/or chemical hydrolysis in the eye (Lee and Li, 1989). These
chemical delivery systems also provide potential to control drug input rate
and prolong drug action in the eye.

Conjunctiva and sclera

The conjunctiva covers most of the ocular surface and has greater
permeability than the cornea, especially for hydrophilic compounds (Wang
et al., 1991). Consequently, depending on dosing conditions and dosage
form, 5–30 times more timolol (Chang and Lee, 1987; Urtti *et al.*, 1990)
and pilocarpine (Lee and Robinson, 1979; Thombre and Himmelstein,

1984; Urtti *et al.*, 1985) is absorbed through the conjunctiva to the systemic circulation than transcorneally into the eye. Conjunctival clearance of the drug from the lacrimal fluid is determined as $CL = P \times S$, where P is the conjunctival permeability (typically in the order of 10^{-5} cm s^{-1}) (Wang *et al.*, 1991) and S is the conjunctival surface area (18 cm^2) (Watsky *et al.*, 1988). Furthermore drug flux through the conjunctiva at steady-state is determined in clearance terms as $J = CL \times dC$, where dC is the concentration gradient of drug between the lacrimal fluid and conjunctiva.

Part of the drug may absorb via conjunctiva to the sclera and thereafter to the ciliary body (Ahmed and Patton, 1985). Drug that is absorbed by this route does not usually gain access to the aqueous humour. Conjunctival and scleral penetration is more favourable than the corneal route, especially for hydrophilic and large molecules like inulin that have poor corneal permeability (Ahmed and Patton, 1985; Ahmed *et al.*, 1987). In addition, conjunctiva has been shown to be the preferable route for ocular absorption of the hydrophilic α_2 agonist *p*-aminoclonidine (Chien *et al.*, 1990).

Drug distribution in the inner eye

After absorption into the aqueous humour a drug may distribute to the surrounding tissues: iris, ciliary body, and lens. A small amount of the drug may also penetrate further to the posterior chamber and vitreous humour (Maurice and Mishima, 1984).

Penetration to the iris and ciliary body takes place easily because these tissues have a porous, leaky surface (Maurice and Mishima, 1984). Consequently, drug concentrations in the iris and ciliary body of albino rabbit reflect those in the aqueous humour readily and they are considered to belong to the same pharmacokinetic compartment (Makoid and Robinson, 1979). Deviations from this relationship take place in the case of pigmented iris and ciliary body if the drug is capable of binding to melanin. For example, pilocarpine (Lee and Robinson, 1982), timolol (Salminen and Urtti, 1984), and atropine (Salazar and Patil, 1976) bind to ocular pigmentation. Typically, binding results in elevated drug concentrations in, and in slower drug elimination from, the tissue (Lee and Robinson, 1982; Salminen and Urtti, 1984). This results in prolonged action of pilocarpine (Urtti *et al.*, 1984) and atropine (Salazar and Patil, 1976) in pigmented eyes compared with less pigmented ones. Thus, pigmentation may behave like an intraocular sustained release depot for some drugs.

The lens is less permeable than iris or ciliary body and, consequently, drug concentrations in this tissue are typically an order of magnitude lower than in the anterior uvea (Maurice and Mishima, 1984; Urtti *et al.*, 1990). The lens decreases drug diffusion from the anterior chamber to the vitreous humour (Maurice and Mishima, 1984). Consequently, higher concentrations are achieved in the back of aphakic than phakic eyes after topical ocular drug administration. Also, convective flow of the aqueous humour from the posterior to the anterior chamber impairs drug access to the posterior eye. Thus, it is not surprising that topical ocular drug administration results in drug concentrations nearly two orders of magnitude lower in the vitreous than in the aqueous humour (Urtti *et al.*, 1990). Consequently, it is difficult to treat disorders in the back of the eye using topical ocular drug administration, while drug treatment of the posterior eye via systemic circulation is impaired by the blood–vitreous barrier (Maurice and Mishima, 1984).

Drug elimination

Drugs are eliminated from the aqueous humour by aqueous drainage through the trabecular meshwork. Drug clearance by this route equals the rate of elimination from the aqueous humour (Conrad and Robinson, 1977; Miller *et al.*, 1981). In addition, many drugs, like pilocarpine and timolol, are eliminated via the blood circulation of the anterior uvea (Miller *et al.*, 1981; Tang-Liu *et al.*, 1984). Consequently, clearance values of pilocarpine, timolol, and flurbiprofen are more than $10 \, \mu l \, min^{-1}$, which is several times higher than the elimination rate from aqueous humour of the rabbit eye (Tang-Liu *et al.*, 1984). In ocular pharmacokinetics it is difficult to distinguish the distribution phenomena from elimination because the lens and vitreous humour can form large and slowly equilibrating drug depots.

After intracameral injection and topical eyedrop application, drug concentration in the aqueous humour decreases according to biphasic kinetics with two half-lives – one for drug distribution and the other for elimination (Makoid and Robinson, 1979; Miller *et al.*, 1981). For example, the apparent half-life for flurbiprofen distribution is 15 min and the half-life for its elimination from the aqueous humour after intracameral injection is 93 min (Tang-Liu *et al.*, 1984).

Although drug absorption from the lacrimal fluid to the cornea ceases in a few minutes, the second step in ocular drug absorption, transfer from the epithelium to the stroma, is slow and may decrease the value of the apparent half-lives of distribution and elimination (Makoid and Robinson, 1979). During the terminal elimination phase the half-life is prolonged by the back-diffusion of the drug from the tissue reservoirs such as the lens, vitreous humour and pigmented uvea (Makoid and Robinson, 1979; Miller *et al.*, 1981).

Subconjunctival drug administration

Antibiotics (Barza *et al.*, 1981) and steroids (McCartney *et al.*, 1965) are examples of drugs which are sometimes administered as subconjunctival injections in order to achieve higher drug concentrations in the eye. The volume administered subconjunctivally is usually 0.5–1.0 ml and after injection a bleb is formed on the ocular surface. Increased ocular absorption of subconjunctival drugs, compared with topically administered drugs, has been demonstrated in several studies (Maurice and Mishima, 1984).

The reasons for increased ocular absorption and the pharmacokinetics of subconjunctival injections have long been studied and discussed but the picture is not yet clear. There are several possibilities for the increased ocular penetration (Fig. 2). Subconjunctivally injected drug may leach from the injection bleb to the lacrimal fluid and thereafter it may absorb transcorneally into the eye (Conrad and Robinson, 1980). Bioavailability is increased because the retention time of subconjunctival depot is longer than retention of eyedrops on the ocular surface. Another possibility is the direct penetration through sclera to the anterior uvea (McCartney *et al.*, 1965). There are studies supporting both views. For example, Conrad and Robinson (1980) have shown in a quantitative and mechanistic study that pilocarpine is absorbed by the rabbit eye mainly via the cornea after subconjunctival injection. In contrast, McCartney *et al.* (1965) have demonstrated scleral penetration of hydrocortisone into the rabbit eye after subconjunctival injection.

The importance of different routes of drug absorption after subconjunctival injection is probably dependent on the physicochemical properties of the drug and the vehicle. Because polar and large molecules do not penetrate well through the cornea (Huang *et al.*, 1983; Wang *et al.*, 1991) they probably do not penetrate corneally even after solution leaching from the subconjunctival depot. For hydrophilic molecules the scleral route may dominate over the corneal route but for more lipophilic molecules leaching from the subconjunctival space and subsequent corneal penetration is the most probable route of ocular absorption. These views are supported by the study of Ahmed and Patton (1985). In their study, inulin (a hydrophilic substance with a molecular weight of 5000) was absorbed after constant topical ocular exposure mainly through conjunctiva and sclera to the iris-ciliary body. In contrast, the smaller and more lipophilic timolol was absorbed mainly transcorneally into the eye. Unfortunately, no one has yet quantitatively and systematically compared ocular absorption of different molecules after subconjunctival injections to demonstrate the possible dependence of the penetration route on the physicochemical properties of the drug.

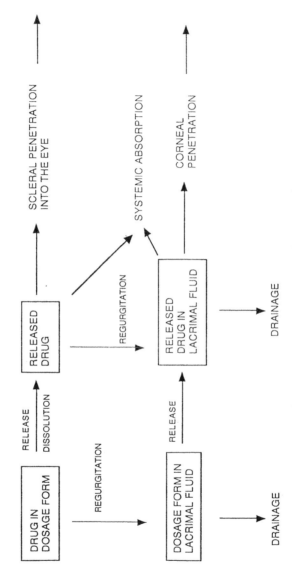

Fig. 2 Ocular pharmacokinetics after subconjunctival injection.

Although at least part of the injected solution may remain in the injection bleb for hours (Conrad and Robinson, 1980) this does not necessarily imply drug retention at the injection site. Subconjunctivally injected drug is usually absorbed rapidly by blood and therefore the drug concentration in the bleb decreases rapidly (Maurice and Mishima, 1984). This happens both in humans and in rabbits (Maurice and Mishima, 1984). For example, pilocarpine concentration decreased in the subconjunctival space of rabbits to 1/100 in 1 h despite a sizeable bleb (Conrad and Robinson, 1980). Furthermore, Maurice and Ota (1978) have shown that a subconjunctivally injected concentration of [^{125}I] iodopyracet decreased tenfold in 30 min in rabbits. In order to increase ocular absorption and prolong the duration of action different subconjunctival prolonged action medications have been tested.

Intravitreal Drug Administration

Drug penetration to the back of the eye is poor due to the anterior barriers – cornea, aqueous turnover, lens, and blood flow in the anterior uvea. Also, penetration of a systemically circulating drug to the vitreous humour is poor. The blood–vitreous barrier consists of the retinal pigment epithelium and the endothelium of the retinal capillaries (Maurice and Mishima, 1984). Consequently, local intravitreal drug administration has gained increasing interest as a possible method for drug administration of antibiotics in endopthalmitis and also in drug treatment of proliferative vitreoretinopathy (Niesman, 1992).

Intravitreally injected drug spreads gradually to the vitreous space. Diffusion in the vitreous is rapid because the collagen concentration in the vitreous gel is low (Maurice and Mishima, 1984). Drugs are eliminated from the vitreous humour via the retina and via the anterior chamber. The rate of drug elimination via the anterior route is determined by molecular weight. The half-life of vitreal elimination for a relatively small molecule such as gentamicin is 1 day, but the half-lives for macromolecules are several days (Maurice and Mishima, 1984). The retinal route contains both passive and active components. Lipid-soluble drugs may be transported across the blood–vitreous barrier and active transport has been described for indomethacin. Both factors may increase the elimination rate so that the half-lives of many antibiotics in the vitreous humour are only a few hours (Maurice and Mishima, 1984).

Sustained release medications have been explored for intravitreal administration because of the rapid drug elimination from the vitreous humour. A second aim is to avoid the retinal drug toxicity that is related to high drug concentrations after intravitreal injection of drug solution.

TOPICAL PHYSICAL DRUG DELIVERY SYSTEMS

In topical ocular controlled drug delivery systems the delivery system releases drug to the tear fluid resulting in prolonged drug concentration in the tear fluid. From the tear fluid part of the drug is absorbed into the eye and the dosing interval of the drug can be prolonged. A prerequisite to the success of this approach is adequate retention of the dosage form in the conjunctival sac (Fig. 1).

Drug release *in vivo* and precorneal kinetics

A topical ocular delivery system for prolonged drug action may be polydisperse (suspension, particulates, liposomes), viscous liquid, gel, gelling vehicle or solid insert (matrix or reservoir device with soluble, erodible or non-erodible polymers) (Shell, 1984; Lee and Robinson, 1986; Gurny *et al.*, 1987). The behaviour of controlled release systems can be substantially changed by physical and chemical characteristics of the polymers or other adjuvants. Polymer properties and other relevant physical and chemical factors offer ways to adjust drug release and dosage form retention in the conjunctival sac. Properties of different polymeric delivery systems in general (Langer, 1980) and in particular for ocular use (Shell, 1984; Lee and Robinson, 1986) have been reviewed. In this paper emphasis will be placed on the kinetics.

During drug release from a prolonged action dosage form its steady-state concentration in the precorneal tear fluid (C_{pc}) is determined as a function of the drug release rate in the preocular area (dQ/dt) and total drug clearance from the precorneal tear fluid (Cl_{pc}):

$$C_{pc} = (dQ/dt)/Cl_{pc} \qquad (1)$$

In this case the volume of the dosage form is constant (e.g. insert, constant amount of microparticulates or suspension crystals after extra solution volume has drained from the eye). Owing to the small preocular volume of the tear fluid ($V_d = 7\,\mu l$) and large precorneal clearance ($12\,\mu l\ min^{-1}$ for timolol) the precorneal system should reach steady-state very rapidly in five half-lives, one half-life being $0.69 \times V_d/Cl_{pc}$ (Gibaldi and Perrier, 1982).

Retention of the dosage form in the conjunctival sac and *in vivo* release rate of the drug are the main factors that determine the drug input (dQ/dt) to the tear fluid after application of the delivery system. Drug release *in vivo* may be similar to *in vitro* release (Urtti *et al.*, 1994) or different (Finne and Urtti, 1992). Poor retention in the preocular area may lead to removal of the dosage form from the conjunctival sac before the drug is released.

As drug clearance from the lacrimal fluid is determined by several parallel elimination pathways (Fig. 1), most notably drainage, tear flow, conjunctival and corneal absorption, drug concentration at steady-state in the lacrimal fluid (C_{pc}) can be determined with a simple equation:

$$C_{pc} = (dQ/dt)/[Cl_{tf} + Cl_{cj} + Cl_{co}]$$

$$= (dQ/dt)/[Cl_{tf} + (P_{cj} \times S_{cj}) + (P_{co} \times S_{co})] \qquad (2)$$

where

dQ/dt = drug release rate *in vivo*,
Cl_{tf} = rate of drug clearance via tear turnover,
Cl_{co} = drug clearance from the lacrimal fluid to the cornea,
Cl_{cj} = drug clearance from the lacrimal fluid to the conjunctiva,
P_{cj} = drug permeability in the conjunctiva,
S_{cj} = conjunctival surface area,
P_{co} = corneal permeability of the drug,
S_{co} = corneal surface area.

In equation 2 clearances across the membranes are calculated as the products of membrane permeabilities (P) and their surface areas (S).

If the dosage form flows partly or completely from the eye after administration, as viscous vehicles, gelling systems, nanoparticulates and liposomes do, it is very important that the drug release rate and the rate of dosage form drainage from the eye match each other (Fig. 1). Too great a drainage rate of the controlled release dosage form from the conjunctival sac may result in poor ocular bioavailability because the drug is not adequately released from the system to the lacrimal fluid before the dosage form is removed. For example, typical liposomes and nanoparticles have half-lives of 3–5 min and 2 min, respectively, on the ocular surface (Fitzgerald *et al.*, 1987a,b). Sometimes drug is not adequately released from liposomes in the tear fluid and, consequently, ocular absorption of liposomal epinephrine (Stratford *et al.*, 1983) and dihydrostreptomycin (Singh and Mezei, 1984) is less than from conventional eyedrops. Release rates of drugs or fluorescent probes from liposomes *in vitro* in buffer solutions are in the order of a few per cent per hour (Fitzgerald *et al.*, 1987a), but *in vivo* release rates may be much higher because the tear fluid increases the rate of drug release from liposomes as much as does plasma (Barber and Shek, 1986). The half-life of carboxyfluorescein release from multi-lamellar vesicles decreases from 6.96 ± 1.03 h to 0.43 ± 0.05 h when the percentage of tear fluid in the release medium increases from 0% to 100%. The mechanism for the increased rate of drug release in lacrimal fluid is not known. Also, in the case of colloid size particles, endocytosis by the conjunctival cells (Latkovic and Nilsson, 1979) may increase the particle retention and drug absorption via conjunctiva and sclera into the eye (Ahmed and Patton, 1986).

Several vehicles have been described that are capable of prolonging drug retention and drug activity in the eye (Shell, 1984; Lee and Robinson, 1986). Viscous, mucoadhesive gels or 'after application' gelling vehicles increase ocular drug absorption by prolonging ocular contact (Lee and Robinson, 1986) but, strictly speaking, these dosage forms are not controlled release systems. They increase both ocular peak drug concentrations and bioavailability, because the drug release from the vehicle is rapid compared with the rate of drug penetration into the eye and, thus, drug release is not the rate-controlling factor in ocular drug delivery (Maurice and Mishima, 1984). This kind of profile has been called prolonged pulse entry of drug into the eye (Salminen et al., 1983).

In contrast to nanoparticles and liposomes, drug release from these systems is faster than their precorneal drainage and drug release is not crucial because the release is rapid. Retention of the vehicle or dosage form on the ocular surface can be misleading. For example, a mucoadhesive hyaluronic acid vehicle shows excellent retention for an eyedrop on the ocular surface (Gurny et al., 1987). Based on gammascintigraphic data, only about 20% of the vehicle had been eliminated from the eye after 10 min whereas the half-life of a saline eyedrop was about 30 s. Despite this order of magnitude difference in the rate of vehicle removal from the eye, ocular biological activity of pilocarpine measured as an area under the miosis versus time curve was improved only 25% and 61% in man and rabbit, respectively, with the hyaluronic acid vehicle. Careful interpretation of the gammascintigraphic data is needed, because the label (usually [^{99m}Tc]-DTPA complex) does not penetrate the ocular tissues. In contrast, pilocarpine readily absorbs (about 24% min^{-1} at a volume of 30 μl) across the conjunctiva to the systemic blood circulation (Fig. 1) (Urtti et al., 1985). Pilocarpine concentration in the tear fluid may decrease much more rapidly than the measured ^{99m}Tc or hyaluronic acid concentration resulting in only modest increases in pilocarpine absorption.

The rate of drug release from some polymeric controlled release systems may be lower in the lacrimal fluid than in in vitro release tests; this may be for several reasons. For example, the mixing conditions in the conjunctival sac are mild and tear volume is very small (7 μl) (Maurice and Mishima, 1984). Monoesters of poly(vinyl methyl ether/maleic anhydride) release timolol at lower rates in vivo in rabbit eyes than in vitro in phosphate buffer (Finne and Urtti, 1992). Because these polymers are polycarboxylic acids their dissolution is dependent on surface pH and transport of hydrogen ions across the static diffusion layer on the polymer surface (Heller et al., 1978).

Consequently, low buffering capacity of the tear fluid and poor mixing result in conditions where both surface pH may decrease and the thickness of the diffusion layer may increase. Consequently, the rate of polymer dissolution and drug release are slowed down *in vivo* (Finne and Urtti, 1992). This factor may affect the drug release from other ionizable and bioerodible polymers in the tear fluid as well.

In contrast to erodible polymer matrices, similar *in vitro* and *in vivo* release rates have been demonstrated for pilocarpine from Ocusert (Urquhart, 1980) and for timolol from silicone reservoir devices (Urtti *et al.*, 1988; Urtti *et al.*, 1994). These polymers neither dissolve nor have interactions with the lacrimal fluid. Here there is no interphase between dissolving solid polymer and already dissolved polymer in the diffusion layer. On the surface of non-erodible reservoir devices there is a diffusion layer with a concentration gradient of the drug but, in most cases, drug concentration in the lacrimal fluid on the polymer surface is substantially lower than its solubility. This hydrodynamic diffusion layer does not limit drug release of timolol from silicone reservoir devices *in vivo* in lacrimal fluid (Urtti *et al.*, 1988, 1994).

In principle, drug release from diffusion controlled delivery systems is determined by the thermodynamic activity gradient of the drug between outer polymer surface and lacrimal fluid. If there is a drug concentration buildup in the receptor medium, drug release rate should decrease because the concentration gradient is decreased. However, owing to the typically high drug clearances from the lacrimal fluid ($5\text{--}10\,\mu l\,min^{-1}$) (Lee and Robinson, 1979; Urtti *et al.*, 1990; Keister *et al.*, 1991) mainly via conjunctival drug absorption, no significant concentration buildup is expected to take place at drug release rates less than $1\,mg\,h^{-1}$ (Urtti, 1991). These calculations are supported by the data on non-erodible timolol inserts of silicone that released timolol at the rates of $4\text{--}7\,\mu g\,h^{-1}$. The release rates *in vitro* and *in vivo* were similar both in rabbit (Urtti *et al.*, 1988) and human (Urtti *et al.*, 1994) eyes. This contrasts with transdermal drug delivery in which poor transdermal drug permeability and subsequent concentration buildup in the occluded sweat may suppress the release rate of the drug from a transdermal patch (Urtti, 1991).

After application of solid inserts, drug concentration in the lacrimal fluid of rabbits is not homogeneously distributed (Urtti *et al.*, 1988). Mixing in the preocular fluid of rabbits is poor owing to the infrequent blinking (four times per minute) and slow tear turnover. This was demonstrated in a study where a silicone device releasing timolol was placed either in the superior or inferior conjunctival sac in pigmented rabbits (Urtti *et al.*, 1988). Application to the superior cul-de-sac resulted in greater corneal timolol absorption compared with the device placement in the inferior conjunctival sac. From the inferior conjunctival sac, timolol absorption took place mainly via the inferior part of the conjunctiva and sclera and thus timolol concentrations in the lower parts of each ocular tissue (cornea, conjunctiva, sclera, iris–ciliary body) were much higher than in the upper parts of the tissues. Poor mixing in the rabbit tear film has been demonstrated also by Lutosky and Maurice (1986). Drug distribution after application of an ocular insert to the human eye may be more even because the blinking frequency is several times per minute (Saettone *et al.*, 1982).

Ocular absorption

Pulse-entry

After eyedrop application drug concentration in the tear fluid decreases rapidly owing to efficient solution drainage and absorption processes (Fig. 1). Consequently, most of the corneal absorption takes place during the first few minutes after instillation of the eyedrop (Chrai *et al.*, 1974).

When viscous and mucoadhesive solutions, gels, or thermosetting gels are used the initial rate of drug removal is slower and more drug may be absorbed by the cornea (Chrai and Robinson, 1974; Saettone *et al.*, 1982; Shell, 1984; Gurny *et al.*, 1987). The possibilities of increasing ocular absorption are, however, dependent on the pharmacokinetic properties of the drug. Very lipophilic drugs rapidly attain their partitioning equilibrium with corneal epithelium and no further increase of drug absorption is achieved by modest prolongation of their corneal contact (Grass and Robinson, 1984; Keister *et al.*, 1991). With a solution of viscosity 90 CP, improvement of the corneal drug absorption was achieved only with water-soluble drugs. Drugs with a log (octanol/pH 7.4 phosphate buffer) distribution coefficient above 1.0 did not show increased ocular absorption when administered in eyedrops with elevated viscosity. Ocular bioavailability is also dependent on the vehicle effects on conjunctival systemic absorption (Ashton *et al.*, 1991).

Without rate control of drug release it is, however, difficult to achieve sustained drug concentrations in the eye. In theory, sustained drug levels and substantial increase in ocular bioavailability can be achieved if the precorneal drug loss rate is $0.1\,h^{-1}$ or less (Maurice and Mishima, 1984). In practice, conjunctival non-productive drug absorption makes this goal impossible. For example, Chang and Lee (1987) blocked the drainage of timolol from the rabbit conjunctival sac mechanically for 2 h, but this resulted in only about a threefold increase in ocular absorption and the apparent half-life of timolol elimination decreased by 15%. In addition, the half-lives of most ocular drugs in the aqueous humour are in the order of

1 h. Consequently, doubling the peak levels extends the concentration curve only by 1 h (Maurice and Mishima, 1984). However, it should be remembered that often the pharmacological activity decays more slowly than the drug concentration in the aqueous humour. This is the case particularly in pigmented eyes and may result in more substantial prolongation of the pharmacological effects through prolonged corneal contact (Urtti *et al.*, 1984).

In conclusion, adequate retention in the conjunctival sac and control of release rate are required in order to achieve sustained drug concentrations, increased duration of activity and decreased peak drug levels in the eye. Otherwise, only limited increase in the duration of activity is achieved, while increased dose and/or improved bioavailability result in the elevation of peak drug concentrations and possible side-effects in the same proportion as absorption was increased.

Controlled release systems

For controlled release systems the fraction of the corneal absorption at steady state is determined as $Cl_c/(Cl_{cj} + Cl_{tf} + Cl_c)$. The equation gives the maximal ocular corneal absorption that can be achieved with the controlled release system staying constantly in the conjunctival sac (Fig. 1). For example, the maximal ocular bioavailability of timolol is 11%, which is approximately 2.5 times higher than the bioavailability after eyedrop administration (Urtti *et al.*, 1990).

Consequently, it is important to minimize the conjunctival permeability relative to the corneal permeability. The corneal drug permeability is more sensitive than the conjunctival permeability to the effect of lipophilicity (Wang *et al.*, 1991). Thus, the fraction of ocular absorption is increased with increasing lipophilicity and also when the drug is administered in a controlled release system. It is obvious that the main factor that limits the bioavailability in the case of controlled release inserts is conjunctival nonproductive drug absorption (Fig. 1). Theoretical calculations by Keister *et al.* (1991) demonstrate that, in the case of very lipophilic drugs, neither increased ocular contact nor decreased eyedrop volume increase the ocular bioavailability. Consequently, the ocular bioavailability of lipophilic drugs is not necessarily increased with ocular controlled release systems, even though the duration of activity and the shape of the response versus time curve is improved. The clearance ratio determines the fraction that is absorbed transcorneally, but the steady-state drug concentration in the aqueous humor is also affected by the rate of drug release (Urtti *et al.*, 1990).

Drug concentration in the aqueous humour is not a good indicator of ocular bioavailability when the drug is absorbed via bulbar conjunctiva and sclera to the eye (Fig. 1). Ophthalmic controlled release systems are usually in contact with the conjunctiva. Conjunctival contact may result in increased drug loss to systemic circulation but conversely it may improve the ocular drug delivery via conjunctiva and sclera. This is an interesting possibility in the ocular delivery of large polar molecules like peptides. For example, selective non-corneal delivery of inulin with liposomes to the iris and ciliary body has been demonstrated by Ahmed and Patton (1986).

Collagen shields

Collagen shields are lens shaped and they are placed on the cornea to protect and to facilitate epithelial healing. Usually the collagen is obtained from porcine sclera. Collagen shields have been studied as potential drug carriers since the late 1980s. Usually, water-soluble drugs are incorporated into the collagen shield by soaking the shield in drug solution. After hydration the water content of the shield is 65–83% (Friedberg et al., 1991). Drug is released after the shield is placed on the cornea. Water-soluble drugs diffuse rapidly from the collagen shields so that their half-life is short, e.g. tobramycin in a 'MediLens' shield has a half-life of only 5 min (Assil et al., 1992). Poorly water-soluble drugs like cyclosporin A are incorporated in the shield during preparation and are released more slowly than more water-soluble drugs (Reidy et al., 1990). Drug release can be modified also by binding the drug to the shield and by changing the degree of cross-linking of the collagen.

Kinetically, collagen shields are exceptional because they are in direct contact with the cornea. Theoretically, the corneal drug absorption relative to conjunctival systemic absorption should be maximized. Gentamicin, tobramycin, vancomycin, dexamethasone, cyclosporin A, and heparin have been administered ocularly in collagen shields (Friedberg et al., 1991). Administration by collagen shields results in increased or comparable ocular drug absorption compared with eyedrops (Friedberg et al., 1991).

There are, however, some potential problems associated with collagen shields. These include mass production problems and high price. In addition, Assil et al. (1992) have shown that collagen shields cause diffuse punctate epitheliopathy which may be one reason for improved corneal drug penetration. These factors may limit the use of collagen shields to selected cases. The most promising area is the postoperative antibiotic treatment of the eyes.

Systemic absorption

After eyedrop administration many ophthalmic drugs are absorbed rapidly and to a large extent by the systemic circulation (Urtti and Salminen, 1993). The main routes of systemic absorption are the nose and conjunctiva (Chang and Lee, 1987).

By using viscous carboxymethyl cellulose vehicle without significant release rate control (Kyyrönen and Urtti, 1990), systemic peak concentrations of timolol were decreased in rabbits 2–3 times, while ocular absorption was increased approximately twofold. These effects may be due to the longer precorneal retention and slower vehicle spread to the nasal mucosa. The effects on peak drug concentrations were more pronounced than the effects on the systemic bioavailabilities (Kyyrönen and Urtti, 1990). In contrast, increased vehicle viscosity did not affect systemic absorption of phenylephrine in monkeys (Kumar et al., 1986).

In the case of polydisperse controlled release systems the systemic drug absorption depends on the drug release and on the vehicle/particulate flow from the conjunctival sac to the nasal mucosa. The faster the dosage form flows to the nose and the longer it stays there the higher the systemic absorption. However, there are only a limited number of studies on the systemic absorption of ocular drugs after administration in polydisperse systems. Compared with eyedrop administration, systemic absorption of triamcinolone acetonide (Singh and Mezei, 1983) and atropine (Meisner et al., 1989) was decreased after ocular application in liposomes.

After ocular administration as solid inserts the systemic absorption kinetics of drugs are very different from eyedrop administration. When drug is administered in a controlled release insert it is released in the conjunctival sac and the systemic absorption may take place mostly through the conjunctiva. For example, timolol clearance from the rabbit lacrimal fluid to the conjunctiva is $10.4 \, \mu l \, min^{-1}$, much greater than the rate of tear turnover, $0.7 \, \mu l \, min^{-1}$ (Urtti et al., 1990). Consequently, only a small fraction of the drug gains access to the lacrimal drainage system. Furthermore, small solution volumes (1–4 μl) do not flow through the lacrimal drainage system to the nose (Lutosky and Maurice, 1986). Consequently, when no extra solution is applied to the eye, drug released from an insert does not necessarily gain access to the nose.

Another kinetic difference between drops and inserts is the rate of drug delivery. Because of the high permeabilities of the conjunctiva and nasal mucosa the systemic drug absorption from eyedrops is often rapid, leading to high and early peak concentrations in plasma (Urtti et al., 1985; Chang and Lee, 1987). By controlling the release, drug input rate to plasma can be controlled and the systemic concentrations decreased (Urtti et al., 1985).

Systemic drug absorption after ocular insert application has been studied only recently. The peak levels of pilocarpine (Urtti *et al.*, 1985), timolol (Finne *et al.*, 1990; Urtti *et al.*, 1990), and morphine (Dumortier *et al.*, 1990) in rabbit plasma decreased substantially with drug application in controlled release inserts instead of eyedrops. The best results were obtained with timolol: compared with an eyedrop, the peak drug levels in plasma were decreased more than 17 times by application in a controlled release silicone insert (Urtti *et al.*, 1990). In humans administration in silicone inserts decreased the systemic peak levels of timolol to about one-third of the levels resulting from eyedrop administration, while equivalent efficacy in reduction of intraocular pressure was achieved (Urtti *et al.*, 1994). Systemic bioavailability is not necessarily decreased with inserts, because the drug is efficiently absorbed through the conjunctiva to the systemic circulation. Despite the decreased peak levels, the systemic bioavailability of timolol was higher after inserts than after eyedrop application (Urtti *et al.*, 1994)

Systemic pharmacokinetics influence the effect that a controlled release system may have on the drug concentration profile in plasma. For example, the peak levels of timolol in plasma decreased by at least 17 times in rabbits and three times in humans when timolol was administered in a silicone device instead of eyedrops (Urtti *et al.*, 1990, 1994). This difference is due to the longer half-life of timolol in humans (4 h) (Benet and Williams, 1990) than in rabbits (0.4 h) (Chang and Lee, 1987). Short half-life kinetics are more sensitive to changes in the drug input rate than long half-life kinetics (Gibaldi and Perrier, 1982). Thus, changes in the release rate are expected to change peak systemic levels of ophthalmic atropine ($t_{1/2}$ about 30 min; Lahdes *et al.*, 1988) more than timolol ($t_{1/2}$ 4 h) in humans.

Drug properties determine whether the systemic bioavailability or peak concentration is more important for induction of side-effects. For example, the systemic side-effects of ophthalmic clonidine are caused by the central effects on α_2-receptors (Hurwitz *et al.*, 1991). Brain is kinetically a deep compartment in which drug concentration does not directly follow the concentrations in plasma; thus, the central effects may not be directly related to drug concentrations in plasma. In this case, the systemic bioavailability is more relevant. In contrast, for β-blockers exerting local direct effects in well-perfused lung and heart, drug levels in plasma may predict the systemic effects. In this case the peak drug concentration in plasma is important (Järvinen and Urtti, 1992). Naturally, the receptor affinities of the drugs vary and thus β_1-selective betaxolol does not cause systemic β blockade at the same concentrations as non-selective timolol (Polansky, 1990).

TOPICAL CHEMICAL DRUG DELIVERY SYSTEMS

In addition to the physical ways of controlling drug release it is possible to control the release rate by chemical reaction. This principle is utilized in prodrugs – compounds that are inactive themselves, but release the parent compound in a chemical reaction, usually enzyme-catalysed hydrolysis. Often prodrugs are esters and esterase activity in the route of drug penetration, or later, is required for activity. In addition to esters, other types of ocular prodrugs have been developed (Lee and Li, 1989).

Ocular prodrugs are usually more lipophilic than the parent compound. Consequently, they show improved corneal absorption. The cornea contains esterases mostly in the epithelium, where hydrolysis of topically applied prodrugs takes place (Lee et al., 1982). Esterases are present also in the corneal stroma and iris–ciliary body and thus they may also have a role in the release of the parent compound. Ocular absorption of epinephrine (Anderson, 1980), timolol (Chang et al., 1987, 1988), and pilocarpine (Mosher et al., 1987; Suhonen et al., 1991) has been improved using the prodrug technique.

Ocular pharmacokinetics of prodrugs is very complex. Although several excellent comparative studies on the general performance of the prodrugs have been undertaken, there are no quantitative modelling studies on ocular prodrugs. The rate and extent at which the parent drug is formed from the prodrug depends on the ability of the prodrug to reach the enzymes, the time that the prodrug is in contact with the enzymes in the cornea, and finally the susceptibility of the prodrug structure to enzymatic hydrolysis (Lee and Li, 1989). The amount of prodrug that is absorbed by the cornea determines the drug bioavailability, if the prodrug is quantitatively converted to the parent drug in the eye. Peak drug concentration in aqueous humour is determined by the input rate and elimination (Chang et al., 1987; Lee and Li, 1989). Here, input rate is determined by the rate of parent drug formation from the prodrug and thus it can be controlled by proper chemical design. Because enzyme-catalysed reactions are saturable, the rate of drug cleavage from the prodrug may be dose-dependent and the ocular pharmacokinetics may become non-linear. Dose-dependent kinetic behaviour of the ocular prodrugs has not been demonstrated but, in principle, it is possible. At the level of enzyme saturation ocular drug delivery could be impaired owing to incomplete prodrug hydrolysis in the eye.

Prodrug technique also offers possibilities of prolonging drug activity in the eye. Prolonged activity is possible if the prodrug is well absorbed into the cornea and the parent drug is released slowly from the prodrug. For example, prolonged miotic activity of lipophilic pilocarpine diester prodrugs was demonstrated by Mosher et al. (1987). Using prodrug technology it is possible to change the rate of hydrolysis by orders of magnitude by changing the ester moieties (Chang *et al.*, 1987; Järvinen *et al.*, 1991). Thus, it is a powerful technique for controlling drug release, but intact prodrug must be retained long enough in the regions of enzymatic activity in the eye (Lee *et al.*, 1982). If the retention in the eye is not long enough to allow drug release from the prodrug, bioavailability may be decreased and no substantial prolongation of drug activity is achieved. Lipophilic drugs with stroma-controlled corneal permeation have longer retention in the corneal epithelium – the main site of prodrug hydrolysis (Suhonen *et al.*, 1991). Conversion of the lipophilic prodrug to less lipophilic intermediates and/or parent drug accelerates the rate of corneal transport (Chien *et al.*, 1991; Suhonen *et al.*, 1991). Thus, it is possible to control the input rate of the drug to the intraocular receptors by changing the rate of hydrolysis and lipophilicity of the prodrug. However, a prerequisite of this is adequate retention of the prodrug in the vicinity of the enzymes in the cornea and iris/ciliary body (Mosher, 1986; Lee and Li, 1989).

Smaller drug doses can be used in prodrug eyedrops because the ocular bioavailability can be increased several-fold (e.g. timolol and adrenaline) (Anderson, 1980; Chang *et al.*, 1987, 1988). Fortunately, the systemic absorption of the drugs does not increase in the same proportion as ocular absorption. Consequently, prodrug technology has been shown to be an efficient way of decreasing the systemic absorption of timolol and adrenaline (Anderson, 1980; Lee and Robinson, 1986; Chang *et al.*, 1987). This follows because the systemic bioavailabilities of timolol and adrenaline are 70% and 50%, respectively, and these values cannot be increased several-fold like ocular absorption.

SUBCONJUNCTIVAL DRUG DELIVERY SYSTEMS

The main kinetic problems of the subconjunctival injections are mechanical drug loss through the injection hole and rapid systemic absorption (Fig. 2). Practical problems are inconvenience and pain.

Compared with the subconjunctival solutions, injected steroid suspensions show different kinetics (Maurice and Mishima, 1984). Prolonged drug concentrations in the eye are reached due to the slow dissolution of the injected steroid crystals in the subconjunctival space. For example, measurable concentrations of hydrocortisone were observed even 7 days after a single subconjunctival injection of hydrocortisone suspension in rabbits. Steroids are very poorly soluble and slow dissolving. For most other drugs, special delivery systems are required to increase the retention time in the subconjunctival bleb and to prolong the drug release.

Sustained release dosage forms may extend the drug retention in the subconjunctival space. This can be achieved by using polymeric inserts. After subconjunctival placement of an insert the free drug concentration in the subconjunctival space determines the driving force of drug absorption from the depot to systemic circulation and to the eye via the scleral route (Fig. 2). For example, biodegradable polylactide-co-glycolide inserts have been used to deliver 5-fluorouracil for 2 weeks after subconjunctival placement (Villain *et al.*, 1992). The application of the system is to increase the success rate of glaucoma filtering surgery (Smith *et al.*, 1992; Villain *et al.*, 1992); 5-fluorouracil is released from the system for 2–3 weeks. In the case of solid inserts mechanical loss from the subconjunctival space is minimal and drug absorption into the eye is determined by the drug release, scleral absorption, and systemic absorption (Fig. 2).

In the case of liposome suspensions mechanical loss from the injection site to the tear fluid is possible. Barza *et al.* (1984) showed that 38% of the lipid remained in the bleb 3 h after the injection. In order to achieve substantial trans-scleral bioavailability drug release from the liposomes should be faster than mechanical loss of lipids from the bleb.

With prolonged action dosage forms it is possible to prevent the rapid systemic absorption of the subconjunctivally injected drugs. Drug molecules bound or encapsulated to liposomes or polymeric insert cannot penetrate to the blood vessels. It should, however, be remembered that once released the drug is absorbed both systemically and ocularly. The parallel non-productive systemic drug absorption is the main factor limiting the ocular bioavailability after administration of subconjunctival sustained release systems.

In practice remarkable prolongation of drug activity in the eye has been observed after subconjunctival administration of inserts and liposome suspensions (Niesman, 1992). Two main applications of the sustained release systems are the administration of antimetabolites (like 5-fluorouracil) to prevent scarring after glaucoma filtering surgery and the delivery of antimicrobial agents to provide prolonged high drug levels in severely infected cornea (Niesman, 1992).

INTRAVITREAL DRUG DELIVERY SYSTEMS

Rapid drug elimination from the vitreous humour is a major kinetic difficulty in intravitreal drug administration. Injection of antiproliferative agents (in proliferative vitreoretinopathy) or antibiotics (in endophthalmitis) may cause retinal toxicity that is associated with the peak drug concentrations (Niesman, 1992). The peak drug concentrations and subsequent toxicity can be decreased by administering the drugs in controlled release systems. Polymeric inserts (Smith *et al.*, 1990), liposomes (Barza *et al.*, 1987; Liu *et al.*, 1989a), and biodegradable microparticles (Moritera *et al.*, 1991) have been used for this purpose.

Liposomes are the most commonly studied prolonged action dosage form for intravitreal administration (Lee *et al.*, 1985). The pharmacokinetics of the intravitreal liposomal drug are presented schematically in Fig. 3. The role of drug release depends on the drug: in the case of antimicrobials only the released fraction is efficacious, while liposomal drug serves as a depot. Consequently, both the efficacy and retinal toxicity of liposomal amphotericin B was decreased compared with the solution (Liu *et al.*, 1989a). This may be due to the fact that after administration only a part of the drug in the vitreous humour is free. Both antimicrobial efficacy and retinal toxicity are related to the free drug concentration and not to the total drug in the vitreous humour. Nevertheless, total drug concentrations (encapsulated plus free) are usually determined from the vitreous humour after liposomal administration. Typically, the half-life of free drug is shorter than that of liposomal drug and, consequently, the elimination and release kinetics of liposomes are the rate-determining factors in the system. Barza *et al.* (1987) determined the half-life of the liposomes in the vitreous humour as 9–20 days, depending on the composition; e.g. the half-life of gentamicin in the vitreous humour is 1 day. Addition of cholesterol to liposomes increases

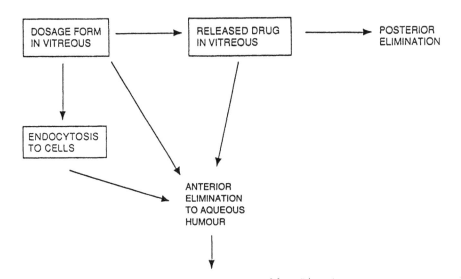

Fig. 3 Ocular pharmacokinetics of intravitreal sustained release medications.

the drug retention in the vitreous humour, probably because the release rate is decreased (Barza *et al.*, 1987).

In the case of antimetabolites, liposomes can be used to target drug to proliferating cells (Stern *et al.*, 1987). The cells may endocytose liposomes and in this way more efficient intracellular drug delivery and drug potency are achieved. Liposomal antimetabolites, e.g. 5-fluorouracil and cytarabine, have reduced retinal toxicity compared with the drug in solution (Stern *et al.*, 1987; Liu *et al.*, 1989b). If the liposomes are endocytosed by the target cells the free drug concentration is not a crucial factor. Three ways of drug elimination from vitreous humour are then possible: elimination of free drug, liposomal drug, and the cells with internalized liposomes (Fig. 3).

After intravitreal liposome injection the concentration of the free drug in the vitreous humour is determined by the drug release from the liposomes and the elimination of the released drug (Fig. 3). Thus, for drugs with high vitreal clearance a faster drug release rate is required to sustain the concentrations at adequate levels. In vitrectomized eyes and for drugs capable of elimination via both the anterior and posterior routes (Maurice and Mishima, 1984) rapid release is required to maintain therapeutic free drug concentrations in the vitreous humour.

Another way to achieve prolonged action for intravitreal drug is to incorporate the drug in biodegradable polylactide-co-glycolide microparticles (Moritera *et al.*, 1991). The microparticles have a mean diameter of 50 μm and therefore cannot be endocytosed (Fig. 3). The microparticles are cleared from the normal eye by 48 ± 5.2 days, but in vitrectomized rabbit eyes the mean retention is 14 ± 2.4 days.

Recently, polymer matrices have been tested as a prolonged action dosage form for intravitreal drug administration. For example, ganciclovir intravitreally administered in sutured ethylene vinyl acetate (EVA) and polyvinyl alcohol polymer matrices resulted in drug levels above the ID100 (inhibitory dose for 100% of the viruses) of cytomegalovirus for 50 days (Smith *et al.*, 1990). This approach may be useful in the treatment of cytomegalovirus infections that are a common problem in the eyes of acquired immune deficiency syndrome (AIDS) patients.

CONCLUSIONS

Prolonged action dosage forms have important benefits in the treatment of several ocular diseases. Together with the rapid development of the controlled release technology and new polymers, better knowledge of ocular pharmacokinetics improves possibilities to develop novel ocular drug delivery systems for topical, subconjunctival and intravitreal use.

ACKNOWLEDGEMENTS

This study was supported by the Academy of Finland.

REFERENCES

Ahmed, I. and Patton, T. F. (1985). *Invest. Ophthalmol. Vis. Sci.* **26**, 584–587.
Ahmed, I. and Patton, T. F. (1986). *Int. J. Pharm.* **34**, 163–167.
Ahmed, I., Gokhale, R. D., Shah, M. V. and Patton, T. F. (1987). *J. Pharm. Sci.* **76**, 583–586.
Anderson, J. A. (1980). *Arch. Ophthalmol.* **98**, 350–353.
Armaly, M. F. and Rao, K. R. (1973). *Invest. Ophthalmol.* **12**, 491–496.
Ashton, P., Podder, S. K. and Lee, V. H. L. (1991). *Pharm. Res.* **8**, 1166–1174.
Assil, K. K., Zarnegar, S. R., Fouraker, B. D. and Schantzlin, D. J. (1992). *Am. J. Ophthalmol.* **113**, 418–423.
Barber, R. F. and Shek, P. N. (1986). *Biochem. Biophys. Acta* **879**, 157–163.
Barza, M., Kane, A. and Baum, J. (1981). *Invest. Ophthalmol. Vis. Sci.* **20**, 509–514.
Barza, M., Baum, J. and Szoka, F. (1984). *Invest. Ophthalmol. Vis. Sci.* **25**, 486–489.
Barza, M., Stuart, M. and Szoka, F. (1987). *Invest. Ophthalmol. Vis. Sci.* **28**, 893–900.
Benet, L. Z. and Williams, R. L. (1990). In *The Pharmacological Basis of Therapeutics*, 8th edn. (Goodman-Gilman, A., Rall, T. W., Nies, A. J. and Taylor, P., eds), pp. 1650–1735. Pergamon Press, New York.
Chang, S. C. and Lee, V. H. L. (1987). *J. Ocul. Pharmacol.* **3**, 159–167.
Chang, S. C., Bundgaard, H., Buur, A. and Lee, V. H. L. (1987). *Invest. Ophthalmol. Vis. Sci.* **28**, 487–491.
Chang, S. C., Bundgaard, H., Buur, A. and Lee, V. H. L. (1988). *Invest. Ophthalmol. Vis. Sci.* **29**, 626–629.
Chavis, R. M., Welham, R. A. N. and Maisey, M. N. (1978). *Arch. Ophthalmol.* **96**, 2066–2068.
Chien, D. S., Hornsy, J. J., Gluchowski, C. and Tang-Liu, D. D. S. (1990). *Curr. Eye Res.* **9**, 1051–1059.
Chien, D. S., Sasaki, H., Bundgaard, H., Buur, A. and Lee, V. H. L. (1991). *Pharm. Res.* **8**, 728–733.
Chrai, S. S. and Robinson, J. R. (1974). *J. Pharm. Sci.* **63**, 1218–1223.
Chrai, S. S., Patton, T. F., Mehta, A. and Robinson, J. R. (1973). *J. Pharm. Sci.* **62**, 1112–1121.
Chrai, S. S., Makoid, M. C., Eriksen, S. P. and Robinson, J. R. (1974). *J. Pharm. Sci.* **63**, 333–338.

Conrad, J. M. and Robinson, J. R. (1977). *J. Pharm. Sci.* **66**, 219-224.

Conrad, J. M. and Robinson, J. R. (1980). *J. Pharm. Sci.* **69**, 875-884.

Conrad, J. M., Reay, W. A., Polcyn, E. and Robinson, J. R. (1978). *J. Parent. Drug. Assoc.* **32**, 149-161.

Doane, M. G., Jensen, A. D. and Dohlman, C. H. (1978). *Am. J. Ophthalmol.* **85**, 383-386.

Dolder, R. and Skinner, F. S. (1983). *Ophthalmica*, pp. 353-424. Wissenschaftliche Verlagsgesellschaft mbH, Stuttgart.

Dumortier, G., Zuber, M., Chast, F., Sandouk, P. and Chaumeil, J. C. (1990). *Int. J. Pharm.* **59**, 1-7.

Finne, U. and Urtti, A. (1992). *Int. J. Pharm.* **84**, 217-222.

Finne, U., Väisänen, V. and Urtti, A. (1990). *Int. J. Pharm.* **65**, 19-27.

Fitzgerald, P., Hadgraft, J. and Wilson, C. G. (1987a). *J. Pharm. Pharmacol.* **39**, 487-490.

Fitzgerald, P., Hadgraft, J., Kreuter, J. and Wilson, C. G. (1987b). *Int. J. Pharm.* **40**, 81-84.

Francouer, M., Ahmed I, Sitek, S. and Patton, T. F. (1983). *Int. J. Pharm.* **16**, 203-213.

Friedberg, M. L., Pleyer, U. and Mondino, B. J. (1991). *Ophthalmol.* **98**, 725-732.

Gibaldi, M. and Perrier, D. (1982). *Pharmacokinetics*, 2nd edn. Marcel Decker Inc., New York.

Grass, G. M. and Robinson, J. R. (1984). *J. Pharm. Sci.* **73**, 1021-1027.

Grass, G. M. and Robinson, J. R. (1988). *J. Pharm. Sci.* **77**, 3-14.

Gurny, R., Ibrahim, H., Aebi, A., Buri, P., Wilson, C. G., Washington, N., Edman, P. and Camber, O. (1987). *J. Control. Rel.* **6**, 367-373.

Havener, W. H. (1983). *Ocular Pharmacology*. The C.V. Mosby Company, St. Louis.

Heller, I., Baker, R. W., Gale, R. M. and Rodin J. O. (1978). *J. Appl. Polym. Sci.* **22**, 1991-2009.

Huang, H. S., Schoenwald, R. D. and Lach, J. L. (1983). *J. Pharm. Sci.* **72**, 1272-1275.

Hui, H. W. and Robinson, J. R. (1985). *Int. J. Pharm.* **26**, 203-213.

Hurwitz, J. J., Maisey, M. N. and Welham, R. A. N. (1975). *Br. J. Ophthalmol.* **59**, 308-312.

Hurwitz, L. M., Kaufman, P. L., Robin, A. L., Weinreb, R. N., Crawford, K. and Shaw, B. (1991). *Drugs* **41**, 514-532.

Ibrahim, H., Bindschaedler, C., Doelker, E., Buri, P. and Gurny, R. (1991). *Int. J. Pharm.* **77**, 211-219.

Junqueira, L. C., Carnero, J. and Long, J. A. (1986). *Basic Histology*, 5th edn. p. 225. Lange Medical Publishers, Los Altos.

Järvinen, K. and Urtti, A. (1992). *Curr. Eye Res.* **11**, 469-473.

Järvinen, T., Suhonen, P., Urtti, A. and Peura, P. (1991). *Int. J. Pharm.* **75**, 259-269.

Keister, J. C., Cooper, E. R. and Missel, P. J. (1991) *J. Pharm. Sci.* **80**, 50-53.

Kumar, V., Schoenwald, R. D. and Barcellos, W. A. (1986). *Arch. Ophthalmol.* **104**, 1189-1191.

Kyyrönen, K. and Urtti, A. (1990). *Invest. Ophthalmol. Vis. Sci.* **31**, 1827–1833.

Lahdes, K., Kaila, T., Huupponen, R., Salminen, L. and Iisalo, E. (1988). *Clin. Pharmacol. Ther.* **44**, 310–314.

Langer, R. (1980). *Chem. Engn. Commun.* **6**, 1–48.

Latkovic, S. and Nilsson, S. E. G. (1979). *Acta Ophthalmol.* **57**, 582–590.

Lee, V. H. L. and Li, V. H. K. (1989). *Adv. Drug Deliv. Rev.* **3**, 1–38.

Lee, V. H. L. and Robinson, J. R. (1979). *J. Pharm. Sci.* **68**, 673–684.

Lee, V. H. L. and Robinson, J. R. (1982). *Int. J. Pharm.* **11**, 155–162.

Lee, V. H. L. and Robinson, J. R. (1986). *J. Ocul. Pharmacol.* **2**, 67–108.

Lee, V. H. L., Morimoto, K. W. and Stratford, R. E. (1982). *Biopharm. Drug Dispos.* **3**, 291–300.

Lee, V. H. L., Urrea, P. T., Smith, R. E. and Schantzlin, D. J. (1985). *Surv. Ophthalmol.* **29**, 335–348.

Liaw, J. H., Rojanasakul, Y. and Robinson, J. R. (1991). *Proc. Intern. Symp. Control. Rel. Bioact. Mater.* **18**, 615–616.

Liu, K. R., Peyman, G. A. and Khoobehi, B. (1989a). *Invest. Ophthalmol. Vis. Sci.* **30**, 1527–1534.

Liu, K. R., Peyman, G. A., She, S. C., Niesman, M. R. and Khoobehi, B. (1989b). *Ophthalmic Surg.* **20**, 358–360.

Loucas, S. P. and Haddad, H. M. (1972). *J. Pharm. Sci.* **61**, 985–986.

Lutosky, S. and Maurice, D. M. (1986). In *The Preocular Tear Film in Health, Disease, and Contact Lens Wear* (Holly, F. J., ed.), pp. 663–669. Lubbock Texas, Dry Eye Institute.

McCartney, H. J., Dryselak, I. O., Gornall, A. G. and Basu, P. N. (1965). *Invest. Ophthalmol.* **4**, 297–303.

Maichuk, Y. F. (1975). *Invest. Ophthalmol.* **14**, 87–90.

Makoid, M. C. and Robinson, J. R. (1979). *J. Pharm. Sci.* **68**, 435–443.

Marchel-Heussler, L., Maricent, L., Hoffman, M., Spittler, J. and Couvreur, P. (1990). *Int. J. Pharm.* **58**, 115–122.

Maurice, D. M. and Mishima, S. (1984). In *Handbook of Experimental Pharmacology, vol. 69, Pharmacology of the Eye*, (Sears, M. L., ed.) pp. 19–116. Springer-Verlag, Berlin-Heidelberg.

Maurice, D. M. and Ota, Y. (1978). *Jpn. J. Ophthalmol.* **22**, 95–100.

Meisner, D., Pringle, J. and Mezei, M. (1989). *Int. J. Pharm.* **55**, 105–113.

Miller, S. C., Himmelstein, K. J. and Patton, T. F. (1981). *J. Pharmacokinet. Biopharm.* **9**, 653–677.

Moritera, T., Ogura, Y., Honda, Y., Wada, R., Hyon, S. and Ikada, Y. (1991). *Invest. Ophthalmol. Vis. Sci.* **32**, 1785–1790.

Mosher, G. L. (1986). *Theoretical and experimental evaluation of pilocarpine prodrugs for ocular delivery*. Ph.D. Thesis, University of Kansas, Lawrence.

Mosher, G. L., Bundgaard, H., Falch, E., Larsen, C. and Mikkelson, T. J. (1987). *Int. J. Pharm.* **39**, 113–120.

Niesman, M. R. (1992). *Crit. Rev. Therap. Drug Carr. Syst.* **9**, 1–38.

Patton, T. F. and Robinson, J. R. (1975). *J. Pharm. Sci.* **64**, 267–271.

Polansky, J. R. (1990). *Int. Ophthalmol. Clin.* **30**, 219–229.

Reidy, J. J., Gebhardt, B. M. and Kaufman, H. E. (1990). *Cornea* **9**, 196–199.

Rojanasakul, Y., Paddock, S. W. and Robinson, J. R. (1990). *Int. J. Pharm.* **61**, 163–172.

Saettone, M. F., Giannaccini, B., Chetoni, P., Galli, G. and Chiellini, E. (1984). *J. Pharm. Pharmacol.* **36**, 229–234.

Saettone, M. F., Giannaccini, B. and Barattini, F. (1982). *Pharm. Acta Helv.* **57**, 47–55.

Salazar, M. and Patil, P. N. (1976). *Invest. Ophthalmol.* **15**, 671–673.

Salminen, L. and Urtti, A. (1984). *Exp. Eye Res.* **38**, 203–206.

Salminen, L., Urtti, A., Kujari, H. and Juslin, M. (1983). *Graefe's Arch. Clin. Exp. Ophthalmol.* **221**, 96–99.

Schoenwald, R. D. and Ward, R. L. (1978). *J. Pharm. Sci.* **67**, 786–788.

Sendelbeck, L., Moore, D. and Urquhart, J. (1975). *Am. J. Ophthalmol.* **80**, 274–283.

Shell, J. W. (1984). *Surv. Ophthalmol.* **29**, 117–128.

Shih, R. L. and Lee, V. H. L. (1990). *J. Ocul. Pharmacol.* **6**, 329–336.

Sieg, J. W. and Robinson, J. R. (1976). *J. Pharm. Sci.* **65**, 1816–1822.

Singh, K. and Mezei, M. (1983). *Int. J. Pharm.* **16**, 339–344.

Singh, K. and Mezei, M. (1984). *Int. J. Pharm.* **19**, 263–266.

Smith, T., Dharma, S., Rafii, M. and Ashton, P. (1992). *Invest. Ophthalmol. Vis. Sci.* **33**, 737.

Smith, T. J., Pearson, P. A., Blandford, D. G., Goins, K. and Ashton, P. (1990). *Proc. Intern. Symp. Control. Rel. Bioact. Mater.* **17**, 470–471.

Stern, W. H., Heath, T., Lewis, G., Guerin, C. J., Erickson, P. A., Lopez, N. G. and Hong, K. (1987). *Invest. Ophthalmol. Vis. Sci.* **28**, 907–910.

Stratford, R. E., Yang, D. C., Redell, M. A. and Lee, V. H. L. (1983). *Int. J. Pharm.* **13**, 263–268.

Suhonen, P., Järvinen, T., Rytkönen, P., Peura, P. and Urtti, A. (1991). *Pharm. Res.* **8**, 1539–1542.

Tang-Liu, D. D. S. and Sandri, R. (1989). *J. Ocular Pharmacol.* **5**, 133–140.

Tang-Liu, D. D. S., Liu, S. S. and Weinkam, R. J. (1984). *J. Pharmacokinet. Biopharm.* **12**, 611–626.

Thombre, A. G. and Himmelstein, K. J. (1984). *J. Pharm. Sci.* **73**, 219–222.

Tonjum, A. M. (1974). *Acta Ophthalmol.* **52**, 650–658.

Urquhart, J. (1980). In *Ophthalmic Drug Delivery Systems* (Robinson, J. R., ed.) pp. 105–118. Am. Pharm. Assn., Washington DC.

Urtti, A. (1991). *Proc. Intern. Symp. Control. Rel. Bioact. Mater.* **18**, 431–432.

Urtti, A. and Salminen, L. (1985). *Acta Ophthalmol.* **63**, 502–506.

Urtti, A. and Salminen, L. (1993). *Surv. Ophthalmol.* **37**, 435–456.

Urtti, A., Salminen, L., Kujari, H. and Jäntti, U. (1984). *Int. J. Pharm.* **19**, 53–61.

Urtti, A., Salminen, L. and Miinalainen, O. (1985). *Int. J. Pharm.* **23**, 147–161.

Urtti, A., Sendo, T., Pipkin, J. D., Rork, G. and Repta, A. J. (1988). *J. Ocul. Pharmacol.* **4**, 335–343.

Urtti, A., Pipkin, J. D., Rork, G., Sendo, T., Finne, U. and Repta, A. J. (1990). *Int. J. Pharm.* **61**, 241–249.

Urtti, A., Rouhiainen, H., Kaila, T. and Saano, V. (1994). *Pharm. Res.* **11**, 1278–1282.

USP DI (1987). *Drug Information for the Health Care Provider*. 7th United States Pharmacopeial Convention Inc., Rockville, MD.

Villain, F., Davis, P., Kiss, K., Cousins, S., Parel, J. M. and Parrish, R. K. (1992). *Invest. Ophthalmol. Vis. Sci.* **33**, 737.

Wang, W., Sasaki, H., Chien, D. S. and Lee, V. H. L. (1991). *Curr. Eye Res.* **10**, 571-579.

Watsky, M. A., Jablonski, M. and Edelhauser, H. F. (1988). *Curr. Eye Res.* **7**, 483-486.

Zaki, I., Fitzgerald, P., Hardy, J. G. and Wilson, C. G. (1986). *J. Pharm. Pharmacol.* **38**, 463-466.

3

COACERVATION-PHASE SEPARATION TECHNOLOGY

J. G. Nairn

Faculty of Pharmacy, University of Toronto, Toronto, Ontario, Canada

DEFINITIONS

The earliest known appearance of the word, phase, was in 1812 (Onions, 1933). An appropriate definition of the word for this chapter is 'a homogeneous, physically distinct, and mechanically separable portion of matter that is present in a non-homogeneous, physical-chemical system and that may be a single compound or a mixture,' (Gove, 1963).

Phase separation is a broad term that may be applied to various processes such as the formation of a solid or liquid phase from a solution. Examples are the crystallization of a salt or the precipitation of a polymer as the result of the removal of some solvent from solution. The new phase is physically distinct and has different properties compared with its solution and can be separated by mechanical means.

The word coacervate was first noted in 1623 (Onions, 1933). This word may be defined in the present context as 'an aggregate of colloidal droplets (as of two hydrophilic sols or of a sol and ions of opposite charge) held together by electrostatic attractive forces,' (Gove, 1963).

Coacervation is a term used to describe the formation of a coacervate related to phase separation and has been applied to the separation of a colloid from a solution into a phase rich in the colloid called the coacervate and the remaining phase which is poor in colloid (Bungenberg de Jong, 1949a). The coacervate has certain properties that distinguish it from the original solution. The coacervate will form a separate liquid layer, but with stirring may form droplets suspended in the polymer-poor phase; furthermore it is usually more viscous, more concentrated and often has the property of binding or adsorbing onto, or engulfing a solid or liquid which

ADVANCES IN PHARMACEUTICAL SCIENCES
ISBN 0–12–032307–9

may be added in to the system.

Coacervation has also been defined as the partial miscibility of two or more optically isotropic liquids, at least one of which is in the colloidal state. It may also be defined as the production by coagulation of a hydrophilic sol of a liquid phase which often appears as viscous drops instead of forming a continuous liquid phase (Considine, 1983). Luzzi (1976) indicated that a solid precipitate is not formed, but a polymer phase consisting of liquid droplets. Bungenberg de Jong (1949a) indicated that simple coacervation is concerned with non-ionized groups on the macromolecule and complex coacervation is concerned with the charges on the macromolecule and the formation of salt bonds. Luzzi (1976) and Deasy (1984a) indicate that simple coacervation usually deals with systems containing only one colloidal solute and complex coacervation involves systems containing more than one colloid, the solvent being water. Arshady (1990a) uses the term coacervation to include the use of both water and non-aqueous systems and describes simple coacervation and complex coacervation in terms of one or two polymers in aqueous systems, respectively. Another definition of a coacervate proposed by Ecanow *et al.* (1990) is that phase most dissimilar to water in its physical chemical properties. Deasy (1984a) has pointed out that the solvent is continuous on both sides of the interface and that the polymer or macromolecule is able to diffuse between the two phases.

Some authors separate the discussion of microencapsulation by coacervation, which includes the use of temperature change or incompatible solvent addition, from other phase separation methods such as the use of emulsification and/or solvent removal to prepare a polymer-rich phase (Kondo, 1979a; Arshady 1990a,b; Watts *et al.*, 1990). Other authors include all these processes in the term coacervation-phase separation to indicate that a new phase, a polymer-rich phase, is being formed (Luzzi, 1976; Deasy, 1984a,b).

In any case, a process is applied to a polymer solution, usually dilute, to reduce solubility in the system to such a degree that appropriate phase separation of the polymer takes place, that is the formation of a polymer-rich phase. This polymer-rich phase may be used to treat pharmaceuticals and other substances. Alternatively, the formation of the colloid-rich phase may result in flocculation and the colloid is present in a higher dispersed state and is usually not satisfactory for encapsulation (Bungenberg de Jong, 1949a).

HISTORY

The term coacervation is derived from the latin acervus, meaning aggregation, and the prefix co to indicate union of colloidal particles. Bungenberg de Jong and Kruyt described the term in 1929 and 1930, as noted by Bungenberg de Jong (1949a). Two chapters, 'Crystallization – Coacervation – Flocculation' and 'Complex Colloidal Systems' describing the early work of coacervation, were prepared by Bungenberg de Jong (1949a,b). The term was used to indicate the formation of colloid rich liquids brought about by various processes which caused phase separation in aqueous systems of macromolecules or colloids in solution.

In 1954, Green and associates of the National Cash Register Company researched the coacervation process using gelatin and gelatin-acacia for commercial purposes. This led to the publication of a number of patents for the preparation of carbonless carbon paper. The result of the coacervation process was the formation of microcapsules containing a colourless dye precursor. The microcapsules were attached to the underside of the paper, and the dye was released upon pressure from pencil or pen and reacted with an acid clay which coated the top surface of the subsequent page; a copy was formed as a result (Deasy, 1984a; Kondo, 1979b).

A brief history of the process using phase separation by emulsification and/or solvent evaporation/extraction is provided by Arshady (1990b), who indicates that early examples include the formation of microspheres/microcapsules of cellulose beads by Wieland and Determan in 1968, and that of dye-loaded polystyrene microcapsules by Vranken and Claeys in 1970. An extensive list of early patents is provided by Kondo (1979a).

The preparation of microcapsules is the single, most important use of coacervation-phase separation because microcapsules are widely used in many industries such as printing, food, aerospace, agriculture, cosmetics, and especially pharmaceuticals. Other methods of microencapsulation have been investigated and are used extensively. The two other principal methods of microencapsulation are chemical processes which include interfacial polymerization and *in situ* polymerization and also mechanical processes which include, for example, air suspension coating and spray drying (Kondo, 1979c; Deasy, 1984c).

REVIEWS

While it is the purpose of this chapter to inform the reader of recent technology related to coacervation-phase separation, it should be pointed out that there are a number of reviews on coacervation-phase separation in both journals and textbooks. Most textbooks with a title containing the word microencapsulation or controlled release include a chapter or two on coacervation and/or phase separation, and usually other chapters dealing with topics such as different methods of preparing microcapsules and various applications. Journal reviews specifically related to coacervation-phase separation include those of Arshady (1990a,b), Madan (1978), Watts *et al.* (1990), Tice and Gilley (1985), Van Oss (1988–89). Journal reviews which include some discussion of coacervation-phase separation include those of Jalil and Nixon (1990a), Thies (1975, 1982), Luzzi (1970), Nixon (1985). Complete chapters on coacervation-phase separation in textbooks are: Kondo (1979a), Bakan (1980), Calanchi and Maccari (1980), Gutcho (1979), Deasy (1984a,b), Donbrow (1992), Fong (1988), Flory (1953), Veis (1970a), Nixon and Harris (1986), Wong and Nixon (1986). Finally, textbooks which include some discussion of coacervation-phase separation have sections prepared by Sparks (1984), Speiser (1976), Luzzi (1976), Hui *et al.* (1987), Kato (1983), Oppenheim (1986), Benoit and Puisieux (1986), Bakan (1986), and Luzzi and Palmieri (1984).

GENERAL PROCESS

A general description of coacervation-phase separation is outlined below. It should be noted that a large number of changes can be made in the process.

Step 1

In order to produce a suitable product such as microcapsules or microspheres by the methods of coacervation-phase separation, it is necessary to select the polymer or macromolecule which will provide the appropriate coating or matrix characteristics desired in the final product, such as an enteric coating or a product to control the release of the drug. The polymer is dissolved in a suitable solvent so that it is usually fully solvated.

Step 2

If it is desired to encapsulate a core such as a drug or chemical, it may be added to the continually stirred polymer solution to form, preferably, a dispersion of the core in the polymer solution. The solvent for the polymer is selected, preferably so that it does not dissolve the core.

Step 3

One of many processes, such as the addition of a non-solvent for the polymer, is used to bring about coacervation-phase separation of the polymer. This process promotes the formation of a new phase 'the coacervate' in a coacervation process. With stirring; coacervate droplets form which encapsulate the core to form microcapsules. It should be noted that most of the solvent used to dissolve the polymer is now the polymer-poor phase and forms the suspending medium in which the core and the coacervate droplets are stirred; this liquid is sometimes called the manufacturing phase. In the case of solvent removal, for example by evaporation, the polymer phase is enriched in polymer and it will eventually deposit on the core. In this system, because the solvent for the polymer has been removed, it may be necessary to provide another liquid, such as liquid paraffin or water which does not evaporate appreciably and is used to suspend the core and the polymer enriched phase.

Step 4

The polymer-rich droplets containing the drug are further desolvated by a process similar to that noted above, or a different process. The polymer may also be hardened by a number of methods such as thermal desolvation or crosslinking to form a product, microcapsules for example, which preferably does not aggregate.

Step 5

The microcapsules are then collected and may be rinsed with an appropriate liquid to remove unwanted solvents and excipients (Bakan, 1980; Deasy, 1984a).

PROCESSES

In order to bring about coacervation-phase separation, a number of processes have been employed. These have been developed and are being improved in order to meet the many requirements of producing microcapsules or microspheres with appropriate characteristics. Some of the factors which must be considered in the process include:

- Solubility of the polymer
- Solubility of the drug
- Heat stability of drug and polymer
- Proper size range of microcapsules or product
- Formation of coacervate or concentrated polymer solution with suitable characteristics such as appropriate viscosity, deposition onto and adherence to the core

• Appropriate shape characteristics
• Sensitivity of enzymes, biological products or drugs to the process

As a result of the characteristics of the polymer, the properties of the core and the desired features of the final product, which is usually microcapsules or microspheres, a number of processes have been developed to effect coacervation-phase separation. The following discussion is an attempt to organize the various processes and is based on the solubility of the polymer, the number of polymers which are expected to form the coating material and the type of process used in the procedure.

Polymer solubility. The solvent used to dissolve the polymer is water or an organic liquid. In order for coacervation-phase separation to take place, it is first necessary to dissolve the polymer in an appropriate solvent so that it can be induced to separate in a more viscous, but still fluid state, in order that it can surround or mix with the core and then be hardened. While most polymers are either soluble in water, such as gelatin, or in an organic solvent, e.g. ethyl acetate, such as ethylcellulose, some polymers, e.g. cellulose acetate phthalate, are soluble in acetone and also in alkaline water.

Number of polymers. In aqueous systems one or two water-soluble polymers are frequently used to form the polymer wall or matrix of the product. Usually, only a single polymer is used to form the wall or the matrix of the microcapsule when the polymer is soluble in the organic liquid. The intentional incorporation of two organic soluble polymers within the same coat or matrix of the microcapsules is infrequent.

It should be noted that polymers may also be used to induce coacervation-phase separation, to improve the process such as minimizing aggregation, or to stabilize an emulsion during the formation of microcapsules, but not be incorporated into the microcapsule coat. Polymers may also be used to gel the interior, that is the core, of liquid-filled microcapsules (Kondo, 1979a).

Processes

I A single wall-forming polymer soluble in water

• addition of a miscible liquid, a non-solvent for the polymer
• addition of a water-soluble salt
• change of temperature
• addition of an incompatible or non-wall-forming polymer
• adjustment of pH
• addition of reacting ions

II Two or more wall-forming polymers soluble in water

• pH adjustment
• dilution and/or temperature change

- addition of an incompatible or non-wall-forming polymer
- three wall-forming polymers

III A single wall-forming polymer soluble in an organic liquid

- addition of a miscible liquid, a non-solvent for the polymer
- change of temperature
- addition of an incompatible or non-wall-forming polymer
- evaporation with a miscible liquid, a non-solvent for the polymer
- evaporation with an immiscible polar liquid, a non-solvent for the polymer
- evaporation or removal with an immiscible organic liquid, a non-solvent for the polymer

IV Two wall-forming polymers soluble in organic liquids
V One wall-forming polymer soluble in water and one soluble in an organic liquid

While emphasis is placed on the above classification, it would appear that the properties of the core material are not important. This is not true as the core, whether a solid, a mixture of solids, a liquid, a solution, a suspension or an emulsion, for example, may alter or inhibit the process of coacervation-phase separation. A suitable process must be selected in order to obtain a satisfactory product in terms of maximum utilization of the core and appropriate characteristics of the product.

In many cases it is not necessary to have a core present – empty microcapsules are formed. In other cases microcapsules are not readily formed unless there is a core present to promote deposition of the polymer. Only a few procedures which do not pertain to pharmacy will be described in order to exemplify different procedures or materials. Coacervation-phase separation is also used extensively in a number of other fields, such as photography, agriculture, food and biology.

I A single wall-forming polymer soluble in water

Addition of a miscible liquid, a non-solvent for the polymer

When ethanol is added to an aqueous solution of gelatin, there is a competition for the water molecules, and some of the water is removed from the gelatin. The partially dehydrated polymer begins to aggregate and a phase rich in gelatin, the coacervate, separates from solution. The addition of excess ethanol causes the formation of a gelatinous mass which is not satisfactory for microencapsulation. As ethanol is added, the coacervate is formed and tends to envelop water-insoluble powders or liquids. Phase diagrams are useful to describe the appropriate concentrations to select for coacervation (Kondo, 1979a, Deasy, 1984a).

As an example, rose oil has been encapsulated by this process above room temperature. A limited amount of ethanol was added, and the temperature was decreased to harden the gelatin microcapsules. After separation the microcapsules were washed with ethanol and dried. A number of polymers other than gelatin may be used, such as agar, pectin, methylcellulose and polyvinyl alcohol, and a number of other hydrophilic organic liquids have been employed – namely acetone, dioxane, isopropanol (Kondo, 1979a). Other modifications of this method include temperature control, pH adjustment and hardening with formalin (Khalil *et al.*, 1968; Nixon *et al.*, 1968). The core material to be encapsulated should have a low solubility in water in order to obtain an appropriate yield of product and may be either liquid or solid.

A useful development of this process is the formation of nanoparticles which are so named because their size is in the nanometer range. Because of their small size, they may be considered as useful drug delivery devices for parenteral purposes. The preparation of nanoparticles involves the treatment of a solution of gelatin containing a suitable surfactant, at appropriate temperature and pH with ethanol, so that the coacervate region is just reached, as indicated by an increase in turbidity. The coacervate is redissolved so that the molecules exist in the 'rolling up region'. To prevent aggregation of the nanoparticles, the mixture is homogenized. The rolled up gelatin can entrap the core material and then the nanoparticles are hardened with an agent such as glutaraldehyde (Marty *et al.*, 1978; Kreuter, 1978).

Addition of a water-soluble salt

Gelatin and other hydrophilic colloids may be treated with various salts such as sodium sulfate which tend to desolvate the colloid, effecting coacervation. This process tends to require a high concentration of salt, 20–30%, which should be removed by treatment with water. Salts with different water-binding capacity according to the Hofmeister series may be used. It is usually necessary to harden the microcapsules by temperature change, pH adjustment or treatment with formaldehyde to obtain a satisfactory product (Khalil *et al.*, 1968; Deasy, 1984a).

Change of temperature

It has been indicated in a number of reviews that temperature change will promote coacervation-phase separation (Madan, 1978; Sparks 1984). The phase separation is believed to be brought about as a result of a decrease in solubility of the polymer. Thus, a decrease in temperature will promote phase separation of gelatin from solution while an increase of temperature will effect a phase separation for methylcellulose, ethyl hydroxyethylcellulose and hydroxy propylcellulose (Sparks, 1984).

Temperature change is more often used in conjunction with other physicochemical factors such as the use of non-solvents and pH adjustment to effect the appropriate coacervation condition. For example, a dispersion of a drug, aspirin, in gelatin was added to mineral oil then phase separation was promoted by a reduction in temperature then isopropyl alcohol was added and the product was hardened with formaldehyde (Paradissis and Parrott, 1968).

Addition of an incompatible or non-wall-forming polymer

The addition of a polymer which has a high affinity for water has been used to induce coacervation-phase separation of the coating polymer, gelatin. Thus, a core was incorporated in a gelatin solution containing 10 to 25% polyethylene glycol to cause phase separation (Kondo, 1979a). Starch has also been noted to induce phase separation when gelatin is used as the wall-forming material (Madan, 1978). A low concentration of the non-wall-forming polymer may also aid in the control of the viscosity of the solution.

Adjustment of pH

There are a number of polymers which are soluble in water and which possess either or both acidic and basic groups in their structure. Thus, an alteration of the pH causes a change in the ionization of the polymer, leading to insolubility. This effects a phase change and under the appropriate conditions, cores, either liquid or solid, suspended in the polymer solution can be encapsulated when the pH is adjusted. The polymer solution containing the suspended core is allowed to drop into a buffered solution and encapsulation takes place; alternatively, the pH of the aqueous polymer solution containing core is slowly changed. The water-insoluble drug sulfadiazine was encapsulated by permitting an alkaline solution of cellulose acetate phthalate containing the drug to drop into a solution of acetic acid (Milovanovic and Nairn, 1986).

A number of polymers such as casein, phthalylated gelatin, a copolymer of methacrylic acid and methylacrylate have been used to encapsulate various core materials such as photographic materials, solvents and oils.

Addition of reacting ions

Cores of biologically active cells in a suspension of a sodium alginate have been encapsulated successfully by permitting the mixture to flow dropwise into a dilute solution of $CaCl_2$. The calcium ions caused immediate gelling of each droplet. The microcapsules were collected and subsequently treated with a polylysine solution to provide a permanent, semi-permeable membrane. In this process the calcium ions react with the alginate ions to produce a colloid-rich phase entrapping the core material (Lim and Moss, 1981).

II Two or more wall-forming polymers soluble in water

pH adjustment

Under suitable conditions, phase separation will occur with a positively charged colloid and a negatively charged colloid in an aqueous medium. This process is usually called complex coacervation. The polymer-rich phase may be used to entrap a core material, if present, and thus produce microcapsules. The most frequent type of this coacervation process is conducted with gelatin and acacia. A 50:50 mixture of a dilute solution of gelatin and gum arabic (acacia) at about 40°C is mixed and the pH is adjusted to 4. This pH is below the isoelectric point of gelatin, and thus it is positively charged while acacia is still negatively charged. The new phase is the coacervate, which has a colloid composition of about 20% and a composition ratio of about 1:1 of the two polymers. The system is cooled, then formaldehyde is added to rigidize the microcapsule. Other polyanionic colloids such as sodium alginate, or polyvinyl methyl ether–maleic anhydride copolymer have been used to form the complex coacervate, although acacia has been the most extensively studied (Kondo, 1979a).

Dilution and/or temperature change

Variations of inducing phase separation of the complex coacervate include not only adjustment of pH, briefly described above, but also dilution of a concentrated solution of gelatin $\geqslant 6\%$ with warm water or by beginning the process at a low temperature (10°C) and increasing the temperature so that the coacervate is formed. The efficient formation of the gelatin–acacia coacervate and its composition depends upon polymer and salt concentration, pH and temperature, each of which must be controlled or adjusted throughout the process (Kondo, 1979a).

Addition of an incompatible or non-wall-forming polymer

Liquid paraffin was encapsulated by first emulsifying it with a solution of acacia and then treating it with an aqueous solution of gelatin containing a non-ionic polymer, polyethylene oxide. The gelatin concentration, the pH and temperature were adjusted for efficient complex coacervation. The use of polyethylene oxide aids in the induction of coacervation over a wider pH range (Jizomoto, 1984).

Three wall-forming polymers

Finally, it should be noted that systems employing three water-soluble polymers have been investigated; for example, gum arabic as the polyanion and gelatin and haemoglobin as the polycations. In this process the pH is lowered to 5 and the coacervate is comprised of haemoglobin and gum arabic. When the pH is lowered to below 4.3, coacervates of gelatin and gum arabic enclose the first coacervate drops. As a result, capsules with two walls are formed (Kondo, 1979a).

III A single wall-forming polymer soluble in an organic liquid

Addition of a miscible liquid, a non-solvent for the polymer

In this process, the polymer is dissolved in an appropriate organic solvent, then a non-solvent, which is miscible with the polymer solvent, is used to induce phase separation. The non-solvent may be organic or an aqueous liquid (Kondo, 1979a). Frequently, low polymer concentrations are used, along with gradual phase separation to promote appropriate encapsulation. This permits the polymer solution to deposit onto the core and at the same time, flow evenly. The formation of high polymer concentrations tends to provide a demixing effect. The demixing effect and coacervation are best explained by phase diagrams. In Fig. 1 the line AKB represents the formation of a new phase. If the original concentration of polymer is on the SD, and heptane is added, a coacervate is formed; however, if the original concentration is on the line DP and heptane is added, then the polymer tends to come out of solution in a form that is not satisfactory for coating (Kondo, 1979a).

An example of this process is the encapsulation of magnesium hydroxide with ethylcellulose, using dichloromethane as the solvent and effecting phase separation with n-hexane as the non-solvent (Kasai and Koishi, 1977). Aqueous solutions of dyes have been encapsulated by this process. The

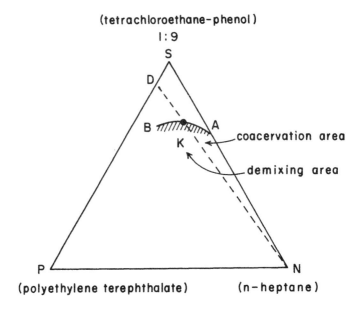

Fig. 1 Ternary diagram showing the phase-separation region of a mixture of polyethylene terephthalate, phenol, and heptane: AB, binodal curve; K, critical point. Reproduced with permission from Kondo (1979a), *Microcapsule Processing and Technology*, p. 96. Marcel Dekker, Inc., New York.

aqueous solution was dispersed in a solution of ethylcellulose dissolved in toluene containing a surface-active agent to assist emulsification. The miscible non-solvent, petroleum ether, was added to induce phase separation of the polymer onto the aqueous droplet (Reyes, 1965). Kondo (1979a) provides an extensive list of polymers, solvents and non-solvents. Some of the polymers are ethylcellulose, cellulose acetate butyrate, polyethylene and polyisobutyl methacrylate. The removal of the residual solvent from the product may be accomplished by washing with more non-solvent or by drying.

Change of temperature

This method of inducing phase separation depends upon the difference in solubility of the coating polymer as the temperature changes. For example, ethylcellulose is dissolved in the solvent, cyclohexane, at about 80°C and as the solution is allowed to cool the coacervate which is formed surrounds the core material which is dispersed throughout the system by appropriate stirring. This process was described by Fanger *et al.* (1970) who reported that after cooling to harden the product, an aggregated product was formed. This system has been studied in detail by Jalsenjak *et al.* (1976). It was found that the procedure was sensitive to minor changes in the procedure. Stirring speed, the vessel geometry and the rate of cooling affected the size distribution of the product, the amount of aggregation, surface characteristics and porosity.

Addition of an incompatible polymer or non-wall-forming polymer

This method of preparing microcapsules by coacervation is similar to the one previously described, in that while temperature is a main factor used to effect phase separation, an incompatible polymer is added to the system in order to aid in the induction of coacervation and/or to minimize the aggregation of microcapsules so that a more uniform product is obtained. As an example of this method, a suspension of the core, aspirin, in a solution of ethylcellulose and polyethylene of low molecular weight in cyclohexane was prepared at 80°C. Upon cooling slowly to room temperature the ethylcellulose separates from solution to surround the core. The polyethylene precipitates from solution in the form of fine particles. The purpose of the polyethylene is not clear, but may aid in the coacervation of the ethylcellulose and also minimize the coalescence of microcapsules prior to hardening of the walls (Kondo, 1979a).

Other incompatible polymers are butyl rubber, polybutadiene, and polydimethylsiloxane. Polyisobutylene, another polymer, apparently acts as a protective colloid and minimizes the formation of agglomerates of ethylcellulose (Deasy, 1984b). Polyisobutylene also promotes the formation of smooth, non-aggregated droplets and it has been shown that it is not incorporated into the wall (Benita and Donbrow, 1980).

After the capsules have been hardened, it is necessary to remove the incompatible polymer. This is readily accomplished, if the polymer remains in solution as the temperature is lowered, by filtration and rinsing the product, or by sieving if the incompatible polymer has a different particle size than the microcapsules.

Evaporation, with a miscible liquid, a non-solvent for the polymer

In this process the polymer is dissolved in an appropriate solvent and then another liquid, the suspending vehicle which has a higher boiling point and which is miscible with the polymer solvent is added. However, this liquid is a non-solvent for the polymer. Prior to evaporation the system is one phase, not including the core which may be added to the polymer solution either before or after mixing with the miscible liquid. During evaporation the polymer separates from solution forming liquid droplets, which are dispersed in the suspending liquid and these coat the core material (Fong, 1988).

An example of this process is the microencapsulation of drug ion-exchange resin complex. The core was dispersed in a solution of polyiso-butylene dissolved in cyclohexane, and light liquid paraffin was added. A solution of ethylcellulose in ethylacetate was then added and evaporation allowed to proceed. The microcapsules were treated with cyclohexane, filtered and washed to remove the suspending liquid (Moldenhauer and Nairn, 1990).

Evaporation with an immiscible polar liquid, a non-solvent for the polymer

The core is dissolved or dispersed in an organic liquid, which contains the dissolved polymer and which has a relatively high vapour pressure and is immiscible with water. This mixture is dispersed in water, the immiscible liquid a non-solvent for the polymer, usually contains surface-active agents or a soluble viscosity agent which aid the formation and stabilization of the resulting oil/water (o/w) emulsion. The organic solvent is removed using heat or by reducing pressure. As the organic solvent is removed, the poly-mer solution becomes concentrated and phase separation of the polymer occurs with the result that the dispersed or dissolved core is entrapped in the polymer matrix (Hui *et al.*, 1987).

As an example of the above, sulfathiazole was dispersed in a solution of ethylcellulose in chloroform. This mixture was then dispersed in an aqueous solution of sodium lauryl sulfate to form an emulsion. After stirring for several hours, the organic solvent evaporated, resulting in the formation of ethylcellulose microcapsules. Other polymers used in this process include polylactic acid, polystyrene, and a large number of hydrophobic polymers (Kondo, 1979a; Deasy, 1984a).

The yield of microencapsulated products is high if the core material has a low solubility in water, otherwise the core will partition into the aqueous phase. The partitioning effect can be decreased if the aqueous phase con-tains salt, which decreases the solubility of the core in the aqueous phase, or by adjusting the pH to decrease the water solubility of the drug. Improved yields of drug can be achieved if the organic solvent has some solubility in water. These solvents then cause rapid deposition of the polymer at the interface, thus forming a barrier that decreases the rate of partitioning of the core into the aqueous phase (Watts *et al.*, 1990).

If the core is an aqueous solution or suspension, it is first dispersed in the polymer solution to give a water/oil (w/o) emulsion and when this is added to the aqueous solution containing the surface-active agent and/or viscosity agent, a water/oil/water (w/o/w) emulsion is formed. The capsule size is influenced by factors such as the viscosity of the starting liquid, agitation speed, and the temperature. It has been suggested that if suitable surfactants are used to prepare the dispersion of the aqueous phase in the polymer solution, small capsules may be prepared with a size of $\approx 10\,\mu m$ (Kondo, 1979a).

An example of this process is the encapsulation of an aqueous solution of an enzyme. This solution is added to a 5 or 10% solution of a polystyrene dissolved in benzene and a primary emulsion is formed by means of a homogenizer. This primary w/o emulsion is then dispersed in an aqueous solution containing a viscosity agent such as gelatin to form the w/o/w emulsion. The temperature is raised to 40°C with constant stirring until the benzene dissolves in the aqueous layer and is removed by evaporation. The polymer is deposited around the aqueous enzyme solution to form the shell wall (Kondo, 1979a).

One difficulty with the process is the time it takes to remove the solvent from the polymer solution, as it is immiscible with the water phase even if the preparation is subjected to heating and reduced pressure. Other techniques used to remove the polymer solvent include freeze drying or adding a solvent that is miscible with water and the polymer solvent but a non-solvent for the polymer (Kondo, 1979a).

A modification of the above process has been called the interfacial deposition technique. In this particular process, n-heptane was emulsified in an aqueous solution of Pluronic F68, an emulsifier, to give an o/w emulsion. A solution of dichloromethane containing either poly (L-lactide) or poly (DL-lactide) was added dropwise to the emulsion which was stirred under partial vacuum. The polymer deposited at the surface of the n-heptane droplets to yield small microcapsules containing water (Makino et al., 1985).

Evaporation or removal with an immiscible organic liquid, a non-solvent for the polymer

This process is especially useful for the preparation of microencapsulated, water-soluble compounds. The drug is dissolved or dispersed in the solution of the organic polymer and this is then dispersed into another organic liquid, usually mineral oil; as a result, an oil/oil (o/o) emulsion or separate phase is formed. The inner phase contains the drug and the polymer and the outer phase is mineral oil. The solvent for the polymer may be partially extracted by the mineral oil and/or may be allowed to evaporate. The resulting microcapsules are filtered and washed with a non-solvent which removes the solvent for the polymer and the mineral oil.

A number of drugs and vaccines have been prepared in microcapsule form in cellulose acetate phthalate in this manner. For example, the drug or vaccine is dispersed in mineral oil, with or without sorbitan monooleate, with stirring and then an acetone-ethanol solution of cellulose acetate phthalate is added. The polymer separates and entraps the drug; after some evaporation has taken place, the mixture is treated with chloroform, filtered and further treated with chloroform (Maharaj et al., 1984; Beyger and Nairn, 1986).

Some investigators, after decanting the excess mineral oil from an ethyl-cellulose ethylacetate system, have placed the product directly in soft gelatin capsules (D'Onofrio et al., 1979).

IV Two wall-forming polymers soluble in organic solvents

This method has been investigated by Itoh et al. (1980) in which a mixture of ethylcellulose and polylactic acid was used to form the wall of the product. The drug sulfamethizole was suspended in the solution of the two polymers which were dissolved in ethylacetate. The miscible non-solvent, pentane, was added dropwise until phase separation occurred and microcapsules were obtained and washed.

V One wall-forming polymer soluble in water and one soluble in an organic liquid

This process, developed by Morris and Warburton (1980, 1982), provides microcapsules that possess three walls. An aqueous acacia solution was dispersed into a solution of polychloroprene in xylene or ethylcellulose in ethylacetate to yield a w/o emulsion. This preparation was then dispersed into an acacia solution to yield a w/o/w emulsion. The organic solvent was allowed to evaporate by bubbling air through the multiple emulsion or removed by dialysis to give microcapsules with a polymeric wall surrounding a solution of acacia. Subsequent evaporation of water leads to the formation of microcapsules with three walls.

CONSIDERATIONS PRIOR TO PROCESSING

The successful preparation of a microencapsulated product depends on the core, the encapsulating polymer and the process of coacervation-phase separation. Once the core has been selected, the formulator will have a choice of a few polymers for a specific pharmaceutical purpose. A consideration of the process must then be made. The selection of the process for the preparation of a suitable product depends upon a number of factors, but principally upon the solubility of the coating polymer and the solubility of the core.

Wall polymer. The polymer used to encapsulate the core is often called the wall polymer and in some cases the polymer will form a wall around a large, solid particle or a liquid drop. However, in other cases the encapsulation polymer may be dispersed throughout the microcapsule providing a matrix type of preparation containing core particles. Finally, the core may be homogeneously or molecularly dispersed throughout the polymer. Almost any non-toxic polymer which is soluble in a solvent can be used to encapsulate pharmaceuticals. Consideration must be given to the intended use of the product such as taste masking, controlled release, enteric coating, biodegradability and modifying the form of a drug, that is from liquid to solid. The polymer should be selected on the basis of its physical and chemical characteristics, its suitability for encapsulating the desired compound and the process which will be used (Luzzi, 1976). A number of properties of polymers have been suggested for the purpose of microencapsulation. These include a cohesive film – a film that adheres to, and is compatible with, the core material and provides the desired properties such as strength, flexibility, impermeability, optical properties, stability, moisture sorption, and solubility. Normal applications require the coat to be from 2 to 30% by weight and this corresponds to a thickness range from 0.1 to 200 μm. However, the amount of coat can vary from 1 to 70% by weight. As noted above, some or a good part of the polymer may be distributed throughout the microcapsule (Bakan, 1980).

Information about appropriate wall polymers may be obtained not only from the literature on coacervation-phase separation, but also from data on cast or free films. These films, however, are not only usually thicker than films enclosing microencapsulated drug particles, but the deposition of cast or free films is considered to be different from the formation of a film by coacervation-phase separation technique, and therefore their properties will usually be somewhat different.

Deasy (1984d) has listed a number of commonly used film formers or polymers for microencapsulation and related uses. The table includes polymers from natural sources, such as gelatin, semi-synthetic polymers such as ethylcellulose, synthetic polymers such as polystyrene and copolymers such as poly(lactic-acid-co-glycolic acid). The release mechanisms of several of these polymers is also given.

Polymer solvents. Once the polymer has been chosen, its solubility characteristics may be determined. The solvent must dissolve the polymer. If the polymer is water soluble, then water may be used to prepare a solution of the polymer; if the polymer is soluble in organic solvents, then a selection of the solvent to be used must be made. Occasionally, a solvent pair will be used. Other considerations for the selection of the organic solvent should be miscibility with another liquid which may be used as a non-solvent, or immiscibility with another liquid, which may be used as the suspending liquid or manufacturing vehicle phase. Furthermore, the polymer solvent should generally not dissolve the core, which may be polar such as water or water-soluble drugs or non-polar such as mineral oil or organic soluble drugs. Other properties to be considered are volatility, toxicity, reaction with the drug and ease of removal from the final product. The polymer-rich phase should have some affinity for the core in order to be adsorbed onto the surface and be fluid enough to encapsulate the core. It is suggested that the process effecting the formation of the polymer-rich phase be slow, and start with a low concentration of polymer. High polymer concentrations generally provide a more rapid phase separation which may be too viscous to provide an appropriate coating of the core. Agglomeration of microcapsules is believed to be caused by the presence of some polymer solvent remaining in the wall. This problem is particularly important when all the solvent is present during the process and until the microcapsules are separated from the suspending liquid. Agglomeration tends to be less when an immiscible liquid is used as the suspending medium because the droplets are appropriately separated in the system. Agglomeration may also minimized by some methods described below (Fong, 1988).

Core. Equal consideration must be given to the core material and its solubility. Types of cores include: (a) no core for the purpose of preparing empty or blank microcapsules; (b) a solid core, e.g., aspirin; (c) a liquid core, e.g., cod liver oil; (d) an active product dissolved in a liquid, e.g., an enzyme in water; (e) on active product dispersed in a liquid, e.g. sulfadiazine in mineral oil; (f) a drug complex, e.g., a drug ion-exchange complex. In order to achieve a high yield of product in terms of active component, it is usually desirable that the core has a low solubility in the polymer solvent or a suspending liquid, used for microencapsulation. However, modifications can be made in the process to improve the yield of the drug product. For example, it may be possible to arrange the conditions in such a way that even if the core material is soluble in the polymer solvent, it has lower solubility than the polymer. In this case the core will come out of solution first and then be encapsulated as phase separation of the polymer takes place. If the above condition is not satisfied, the process of encapsulation will be inefficient and the product will consist of non-encapsulated core and empty microcapsules and microcapsules containing some core (Fong, 1988).

Solid cores are used more often than liquid cores. One reason for this is that the particle size may be more readily controlled. Large-size cores frequently have only one core particle per microcapsule (Moldenhauer and Nairn, 1990) providing a membrane type of microcapsule. Cores of smaller size may have several particles per microcapsule, either due to agglomeration of core particles before microencapsulation, or as a result of the formation of aggregates of coacervate droplets containing core particles. This leads to a matrix type of microcapsule. Finally, if the core dissolves in the polymer solvent, then provided crystallization or precipitation of the core does not take place, the microcapsule may contain a molecular dispersion of the core in the polymer phase. The shape of the core particle frequently will control the shape of the final product, particularly if the core to coat ratio is high. Thus, grinding, milling and spheronization may be considered in order to obtain the appropriate core size and shape. In addition, polymorphs of a drug may be considered in order to obtain a shape that is more amenable to encapsulation (Deasy, 1984d).

Wall-forming polymers soluble in water

Wall polymer. Wall polymers include acacia, alginate, carboxymethylcellulose, gelatin, polyethylene glycol, poly (vinyl alcohol), albumin, carbupol and pectin. Appropriate polymers may be used singly or in pairs as described below.

Cores – water-soluble. In general, water-soluble solids or liquids are not encapsulated to a great extent when water-soluble polymers are used because the core will be distributed between the aqueous polymer-rich phase and the aqueous polymer-poor phase. There are, however, techniques that may be used to encapsulate water-soluble compounds with water-soluble polymers (Harris, 1981).

1. Make the water-soluble core such as KCl insoluble in water through the use of waxes such as carnauba.
2. While the control of pH is important for water-soluble polymers such as gelatin, it also may be used to alter the solubility of many drugs which are weak acids or bases. Thus, the solubility of salicylic acid is decreased in acid solution and may be a candidate for encapsulation by water-soluble polymers.
3. Preparing water-insoluble cores by first making microcapsules using a water-insoluble polymer and then providing a second coat with a water-soluble coating.

Cores – solid. The encapsulation of a number of solid core materials by means of simple coacervation has been studied in an organized manner by Okada *et al.* (1985a). The ability of different core particles to be encapsulated with gelatin was studied as a function of different miscible non-solvents, and other manufacturing parameters. The effect of solubility, zeta potential and the adsorption of gelatin was related to the ability of the product to be encapsulated. Low solubility, high gelatin adsorption and zeta potential play a significant role in the ability of the process to encapsulate the core.

Cores – liquid. Research has also been carried out on the encapsulation of liquids with water-soluble polymers. Gelatin–acacia microcapsules containing oils or oils containing a drug were prepared by employing polyethylene glycol or polyethylene oxide as the incompatible or non-wall-forming polymer. After cooling, the microcapsules formed were cross-linked with glutaraldehyde (Jizomoto, 1984). Jizomoto also showed that the minimum concentration of the polymer polyethylene glycol (or polyethylene oxide) necessary for complex coacervation depended upon the molecular weight. The molecular weight may be related to the chemical potential and the excluded volume of the polymer used to effect coacervation (Jizomoto, 1985).

Process variables. The total polymer concentration has been shown to be related to droplet size. The mean diameter of the coacervate droplets of gelatin–Carbopol 941 microspheres increased from 50 μm to 135 μm as the concentration of polymers increased fivefold (El Gindy and El Egakey, 1981a,b). Similar results have been described by Mortada *et al.* (1987a) for the gelatin–Gantrez system.

El Gindy and El Egakey (1981a) showed that the droplet size decreased as the speed of rotation increased for the gelatin–Carbopol system; at a speed of 600–650 r.p.m. the mode was about 35 μm, while at 100–150 r.p.m. the mode was approximately 85 μm.

In order to minimize aggregation in the complex coacervate system, gelatin–acacia, Maierson (1969) added a surfactant to the prepared microcapsules, and as a result of steric and charge effects, microcapsules were kept apart and aggregation minimized.

A single wall-forming polymer soluble in an organic liquid

Addition of a miscible liquid, a non-solvent for the polymer

Wall polymer. Wall-forming polymers include acrylates, cellulose acetate, cellulose acetate butyrate, ethylcellulose, poly(lactic acid), poly(lactic-co-glycolide), polystyrene, polyvinyl acetate, polyvinyl chloride and other polymers (Fong, 1988).

Control of agglomeration. Fong (1988) described a number of patents regarding agglomeration. Agglomeration of poly(lactic acid) microcapsules prepared from a dispersion of drug particles in a solution of the polymer using a non-solvent was minimized by conducting the phase separation at a temperature of −65°C using a dry ice bath. The use of low temperatures made the wall of the microcapsules sufficiently firm during the phase separation process, so that adhesion of the microcapsules was avoided. Another technique for minimizing the agglomeration of microcapsules employs talc. During the addition of the non-solvent, talc, a mineral silicate is added to minimize the adhesion and coalescence of the microcapsules. It is suggested that the talc forms a barrier against adhesion between the microcapsules. Talc has been used to minimize agglomeration in other patents. Polyisobutylene has been used to minimize aggregation of microcapsules prepared from Eudragit RS (Chun and Shin, 1988) and from Eudragit RS100 (Chattaraj *et al.*, 1991).

Polymer solvent. The polymer solvent must be miscible with the non-solvent and should not dissolve the core. The choice of a solvent for a particular polymer can have an effect on the product and its properties. A methacrylate polymer, Eudragit RLPM, has been used to encapsulate riboflavin. When the polymer is dissolved in benzene and then treated with petroleum ether as the non-solvent, phase separation occurs, with the result that large polymeric droplets are formed which adsorb on the surface of the vitamin as a thick, uniform coat. The product provides a slow release of riboflavin. This is in contrast to the use of isopropanol as the solvent which, after treatment with non-solvent, provides smaller coacervate droplets, a thinner coat and faster release (El Sayed *et al.*, 1982).

Non-solvent. The miscible non-solvent should effect phase separation of the polymer and not dissolve the core material. The non-solvent should be easily removed by evaporation or by rinsing with a volatile solvent possessing similar properties to the non-solvent. Both polar and non-polar non-solvents have been used in the formation of microcapsules. Fong (1988) preferred polar non-solvents such as isopropanol and isobutanol to non-polar, non-solvents such as heptane in low temperature microencapsulation. He found that a combination of non-solvents, e.g. propylene glycol and isopropanol, produces larger microcapsules (100–125 μm) than those prepared from isopropanol alone (25–50 μm). In some cases, it is easier to control the wall thickness when the appropriate non-solvent is used to prepare polystyrene microcapsules (Iso *et al.*, 1985a,b).

Core. Generally, the core should be insoluble in both the solvent for the polymer and the non-solvent. A wide variety of cores have been encapsulated by this method, including antibacterial and anticancer agents, steroids, vitamins, antacids, and pharmaceuticals (Fong, 1988). Products with two walls have also been encapsulated (Hiestand, 1966). Methods have been used to encapsulate cores which are soluble in the solvent. For example, soluble thioridazine as the free base was soluble in the polymer solvent, but after converting to the pamoate salt it was insoluble in both the polymer solvent and the non-solvent. Another method for encapsulating a soluble core is to begin phase separation of the polymer before adding the core particles. As an example, after a solution of styrene maleic acid copolymer in ethanol was prepared, the non-solvent isopropyl ether was added until turbidity was first noticed. The drug methylprednisone was then added and the rest of the isopropyl ether was added to complete the process (Fong, 1988).

The non-solvent method of inducing phase separation has been used to prepare products which have two polymers to alter the release of the core. Itoh and Nakano (1980) coated matrix particles composed of an evaporated product of drug and cellulose acetate with ethylcellulose. In a patent, Fong (1988) describes the preparation of microcapsules with a double wall of polylactic acid prepared by essentially repeating the process.

Change of temparature

Wall polymer. The polymer selected for this process must have a low solubility at room temperature and a high solubility at elevated temperature where it is completely dissolved. Few polymers possess this property, for example, ethylcellulose, hydroxyethylcellulose, hydroxypropylmethylcellulose, methylcellulose (Fong, 1988). The molecular weight of the ethylcellulose affects some of the properties of the final product. Deasy *et al.* (1980) showed that a higher molecular weight of ethylcellulose (100 cp, 0.1 Pa s) gave finer microcapsules and slower drug dissolution than those microcapsules prepared with lower molecular weight ethylcellulose (10 cp, 0.01 Pa s).

Control of agglomeration. Koida *et al.* (1983) found that the agglomeration of microcapsules was affected by the molecular weight. For example, microcapsules in the 149–250 μm size range increased as the molecular weight of the ethylcellulose increased. Agglomeration of microcapsules can be minimized by vigorous agitation, slow cooling near the coacervation temperature and washing with cold solvent (Deasy *et al.*, 1980). Several aliphatic solvents at low temperature, 10°C, have been used to minimize aggregation: pentane, hexane or octane and also cyclohexane (Morse *et al.*, 1978). The addition of a polymer for the purpose of minimizing aggregation has been investigated. Both butyl rubber and polyethylene have been investigated, however, polyisobutylene has been studied more intensely. Samejima *et al.* (1982) indicated that polyisobutylene prevented aggregation of microcapsules and found it to be equally as effective as butyl rubber and much better than polyethylene. Donbrow and Benita (1977) and Benita and Donbrow (1980) indicated the beneficial effects of polyisobutylene in preventing aggregation and suggested that it was adsorbed onto the surface of the ethylcellulose droplet, probably functioning as a protective colloid. Ethylene vinyl acetate copolymer has also been shown to alter the particle size and perhaps the aggregation of ethylcellulose microcapsules (Lin *et al.*, 1985).

Control of particle size. The particle size of the coacervate drops was shown to decrease as the viscosity of the medium increased as a result of higher concentrations of polyisobutylene in the preparation of ethylcellulose microcapsules in cyclohexane by temperature change (Benita and Donbrow, 1980).

Core. Fong (1988) has provided a list of cores that have been encapsulated by this method; these include aspirin, vitamin C, sodium salicylate, chloramphenicol, isoniazid, phenethicillin potassium, sodium phenobarbital, niacinamide and riboflavin. The core compounds should have a low solubility in the solvent at the coacervation temperature. In addition, the compounds should be stable at the temperature employed. Koida *et al.* (1986) found that the efficacy of encapsulation was also related to low solubility of the core material.

Addition of an incompatible or non-wall-forming polymer

Incompatible polymers. The incompatible polymer is chosen on the basis of its higher solubility in the solvent than the coating polymer, thus there will be a tendency for the coating polymer to come out of solution first and coat the core, resulting in a core that is surrounded by one polymer only. Polymers that have been used are frequently low molecular weight liquids, mainly polybutadiene, methacrylic polymer and polydimethyl siloxane. It has been indicated that the advantage of using an incompatible polymer is that certain properties of the coacervation phase, namely the viscosity and relative volume, can be controlled. If the viscosity of the coacervation phase is too high, proper coating of the core cannot occur (Fong, 1988). Extensive studies on the system poly(DL-lactic-co-glycolic acid) copolymer dissolved in methylene chloride using silicone oil have been made by Ruiz *et al.* (1989).

Wall polymers. Some polymers used to form the walls are ethylcellulose, polymethylmethacrylate, polystyrene and poly (lactic-co-glycolide). The wall material may be hardened by adding a non-solvent for the polymer, for example microcapsules of methylene blue hydrochloride prepared from the wall polymer ethylcellulose dissolved in toluene employing the incompatible polymer polybutadiene were hardened by treatment with hexane. Other techniques include solvent evaporation and chemical cross-linking (Fong, 1988).

Control of agglomeration. It has been suggested that agglomeration can be minimized when solvent evaporation is used to harden the wall by using excess liquid paraffin and/or low temperatures during the microencapsulation process (Fong, 1988).

Solvent. The solvent must dissolve the wall polymer, the incompatible polymer and it should also be miscible with the washing non-solvent and should not dissolve the core.

Washing solvent. The function of the washing solvent is to remove the polymer solvent from the microcapsules and any incompatible polymer. Consequently, the washing solvent should be miscible with the polymer solvent, dissolve the incompatible polymer and not dissolve the core material or the polymer wall material.

Core material. Some examples of core materials encapsulated by this method include antibiotics, pharmaceuticals, polypeptides and dyes (Fong, 1988).

Evaporation with a miscible liquid, a non-solvent for the polymer

Wall polymer. Polymers include primarily ethylcellulose, polyethylene, ethylene acrylic copolymers and vinyl polymers (Fong, 1988).

Polymer solvent. The polymer solvent should have a relatively high vapour pressure so that it is readily evaporated. Polymer solvents include aliphatic and aromatic hydrocarbons, ketones, ethers, alcohols and esters.

Miscible liquid, a non-solvent for the polymer. The miscible liquid should not dissolve the core or the polymer, but should be miscible in the concentration used with the polymer and solvent, thus at the beginning of the process the mixture is homogeneous. Furthermore, this liquid should have a low vapour pressure so that during evaporation it will function as the suspending liquid. Examples of suspending liquids include hydrocarbons with a high boiling point, silicone fluid and polyethylene glycols (Fong, 1988).

Control of agglomeration. Moldenhauer and Nairn (1990) indicated that polyisobutylene has other effects, in addition to its protective colloidal action, namely, increasing the viscosity of the system, thereby suspending the core more uniformly, especially larger particle sizes, and decreasing the rate of evaporation, thereby the rate of surface nucleation of the polymer. It has also been suggested that the use of the suspending medium, light liquid paraffin, in this case actually initiates the coacervation process by removing some of the solvent from the polymer.

Coat structure. The rate of evaporation has an effect on the coat structure and thereby the rate of release of the microencapsulated drug (Moldenhauer and Nairn, 1991). Intermediate rates of evaporation provide a dense coat of uniform thickness and a smooth surface, whereas fast evaporation rates produce an irregular, smooth, porous coat and slow rates of evaporation produce a sponge-like coat.

Evaporation with an immiscible polar liquid, a non-solvent for the polymer

Core. The most important factor in selecting a core material is that it has a low solubility in the polar liquid, usually water, otherwise some of the core will partition into the aqueous external phase. A number of cores have been encapsulated, as listed by Fong (1988), such as antibiotics, antineo-plastics, anaesthetics, insulin, steroids and other pharmaceuticals. Water-soluble drugs are generally not successfully encapsulated by this method; for example, salicylic acid, theophylline or caffeine could not be encapsulated with polylactic acid from a preparation of the drug in methylene chloride (Bodmeier and McGinity, 1987a).

Several methods have been used to encapsulate water-soluble or partially water-soluble drugs. Weak bases or weak acids in their salt form may be converted to their non-ionic form, thereby reducing their solubility. Chemical modification of a compound can be used to reduce its water solubility, thereby making it easier to encapsulate by this method. The addition of an inorganic salt to the aqueous phase will reduce the solubility of the core. Alternatively, some of the core can be added to the external aqueous phase in order to decrease the partition of the drug from the core to the external aqueous phase. For example, the addition of tetracaine to the non-solvent more than doubled the drug content of the microspheres (Wakiyama *et al.*, 1981). Other examples of loading in the external aqueous phase with drug in order to obtain a high drug content in the microcapsule include a saturated solution of quinidine sulfate (Bodmeier and McGinity, 1987b), and cisplatin (Spenlehauer *et al.*, 1988).

The solubility of the core in the solvent for the polymer will have an effect on the nature of the final product. If the core material is soluble in the polymer solvent, then the encapsulated product will tend to have a homo-geneous structure, as both the core and the polymer will come out of solution as the solvent is evaporated. If the core material is insoluble in the polymer solution, then thought should be given to the particle size before beginning the encapsulation process and milling or micronization should be considered. When the polymer comes out of solution, it will surround the core particles, leading to a heterogeneous product. Finally, large cores have been encapsulated by this method with the result that a membrane covers the drug. This type of product is different from the two types described above which are either homogeneous or heterogeneous in nature. The rate of release from the single core will tend to be constant while that from the monolithic type of microcapsule will tend to decrease with time. Blank microcapsules, that is microcapsules without a core, may also be prepared (Fong, 1988).

The core loading will affect the ratio of polymer to core, the size of the microcapsule, and the rate of release of the core material. The rate of release of dibucaine (Wakiyama *et al.*, 1982), butamben, tetracaine (Wakiyama *et al.*, 1981), ketotifen, and hydrocortisone acetate (Fong, 1988) all increase as the core loading increases. The maximum fraction of core loading depends upon the properties of the microencapsulation system, but may range from 0.4 to 0.75; for example, thioridazine and ketotifen were encapsulated at a fraction of 0.5 to 0.6 (Fong, 1988).

As mentioned in the process section, aqueous solutions have been encapsulated with considerable success, leading to a w/o/w system; however, a water-soluble compound will tend to diffuse into the outer aqueous phase. The loss of water to the external aqueous phase is reduced by using humectants such as glycerin (Fong, 1988). Gelatin has also been used as an internal stabilizer (Kondo, 1979a).

Wall polymers. The polymer selected for this process must be insoluble in water. Some examples include ethylcellulose, polystyrene and cellulose acetate butyrate which are used to prepare microcapsules. A number of biodegradable polymers have been used to prepare microspheres of pharmaceuticals; these include homopolymers and copolymers of lactic acid, glycolic acid, β-hydroxybutyric acid and caprolactam (Fong, 1988). This phenomenon occurs with poly(DL-lactide) (Spenlehauer *et al.*, 1986). Generally, particle size increases with polymer concentration.

The concentration of the wall-forming polymer solution, that is the polymer to solvent ratio, has an effect on the *in vitro* release rate of the core material. The release of the core material decreased when the initial concentration of the polymer solution was increased. The significance of this factor depends upon the drug that is used as the core. When thioridazine was used as the core material, the effect was appreciable as the initial polymer concentration in the process was raised from 5 to 10%.

However, only a small change was noticed when the core was hydrocortisone acetate. It has been suggested that the formation of homogeneous microcapsules containing thioridazine, which is soluble in the polymer solution, was more readily affected by the initial polymer concentration in the solvent than for the hydrocortisone acetate which was not soluble in the polymer solution, and thus formed heterogeneous microcapsules (Fong *et al.*, 1986).

Polymer solvents. The solvent for the wall-forming polymer should be immiscible or have only a low solubility in water. Its boiling point must be lower than that of water so that it will evaporate faster than the external phase water. A solvent frequently used in this microencapsulation process is methylene chloride because of its high vapour pressure and because it is a good solvent for many polymers. Methylene chloride is, however, toxic and considerable amounts can remain in the product even after drying. Weight losses of up to 3.5% were determined by thermogravimetric analysis and chlorine content analyses (Benoit *et al.*, 1986). Other solvents for polymers include chloroform, carbon tetrachloride, ethylene chloride, ethyl ether, benzene and methyl acetate (Fong, 1988).

Aqueous phase. The solubility of the polymer solvent in the aqueous phase has been shown to have a significant effect on drug loading. A study of solvent effects on the entrapment of quinidine sulfate showed that high loading was achieved with the solvent methylene chloride, which had the greatest water solubility, whereas very poor loading was achieved using chloroform, which has a lower water solubility. It is suggested that solvents with high water solubility effect rapid deposition of polymer at the droplet interface, creating a barrier at the interface, thus minimizing drug diffusion out of the microsphere to the outer phase water. Alternatively, if water-miscible polymer solvents are used to dissolve the drug and polymer, agglomerates of polymer are formed on mixing. Mixtures of polymer solvents with different water solubilities can be used to obtain microspheres (Bodmeier and McGinity, 1988).

Surfactants and emulsifying agents. The emulsifying agent should be selected so that an emulsion of the appropriate particle size is readily formed, and it stabilizes the emulsion during removal of the volatile polymer solvent preventing coalescence of the droplets. As this method of phase separation involves the formation of an o/w emulsion, the proper HLB (hydrophile-lipophile balance) value is 8 to 18. The specific emulsifying agent and its concentration should be determined by trial and error.

Salts of fatty acids, particularly sodium or potassium oleate, have been found to be useful for the preparation of microcapsules, including polymers that are subject to biodegradation. As an example, sodium oleate produced high yields of biodegradable microspheres which were free of agglomeration. The fraction of the drug incorporated was 80–99%, and core loading was up to 0.5 of the weight of the microsphere. The size of the microspheres was less than 150 μm diameter, which will pass through a 20 gauge needle (Fong *et al.*, 1986). It is necessary to consider the pH of the aqueous phase if the surface active agents are subject to different degrees of ionization as a result of different pK values. It will be necessary to maintain a pH at least two to three units above the pK_a of the fatty acid in order for it to be properly ionized.

Surface-active agents, both anionic and non-ionic with an HLB of at least 10 at a concentration of 0.1–1% have been used to prepare microcapsules by this method (Morishita *et al.*, 1976). Watts *et al.* (1990) have listed a number of surface-active agents: polysorbate 80, sodium oleate and sodium dodecyl sulfate. The use of polysorbate 80 in the aqueous phase at a concentration of 2% produced a small reduction in the content of quinidine in the microcapsules. This was attributed to not only an increased solubility of the drug in the aqueous phase, but also to stabilization of the polymer droplet interface which reduced the rate of solvent loss, thereby reducing the polymer deposition rate and permitting a greater loss of drug from the partially formed microcapsules before a suitable hardened barrier could be formed (Bodmeier and McGinity, 1987b). Low yields of small microcapsules may be obtained if surface-active agents are used, e.g. sodium lauryl sulfate (Jaffe, 1981).

Emulsifiers such as gelatin and polyvinyl alcohol at a concentration of 0.5–2.0% may be used to form o/w emulsions (Morishita *et al.*, 1976). Emulsifiers have other effects on the preparation of microspheres as a result of enhanced solubilization. Lomustine and progesterone crystals were formed on the microsphere surface and in the aqueous phase as a result of using polyvinyl alcohol and methylcellulose. The crystals were eliminated and drug loading improved when the emulsifier was removed half way through the evaporation step (Benita *et al.*, 1984). The use of emulsifiers can alter the rate of release of drugs from the microsphere. For example, the use of a gelatin solution which provides a lower solubility for insulin showed only a 26% burst effect, compared with a solution of polyvinyl alcohol which provides a higher solubility for the insulin, resulting in an 88% burst effect. The difference in the burst effect has been attributed to the difference in the solubility of insulin in the hydrophilic colloidal solution (Kwong *et al.*, 1986).

Wakiyama *et al.* (1982) investigated the effect of acid-processed gelatin and alkaline-processed gelatin as an emulsifying agent on the yield and efficiency of microencapsulation of basic amino drugs. A greater efficiency of drug incorporation was achieved when alkaline-processed gelatin was used, perhaps owing to the fact that alkaline-processed gelatin gave a pH of 7.5, promoting the formation of the non-ionized form of dibucaine (pK_a 1.6 and 8.3) and thus its greater uptake by the solvent as a result of the greater o/w partition coefficient. When the pH of the aqueous phase was raised to 8.6, a greater fraction of the dibucaine was in the non-ionized form and a greater incorporation of drug was observed.

The use of hydrophilic colloids to stabilize the emulsion may have an effect on the shape or particle size of the final product. For example, the use of methylcellulose 400 and polyvinyl alcohol as a stabilizer for the external aqueous phase resulted in oval-shaped microcapsules; most of the products were, however, spherical (Cavalier et al., 1986). In other experiments microsphere size was dependent on the type and concentration of the emulsifying agent; for example, microsphere size increased with increasing polyvinyl alcohol concentration (Benita et al. 1984). Other researchers have obtained smaller microspheres with a 1% sodium alginate solution which had a higher viscosity than a 1 or 2% gelatin solution (Wakiyama et al., 1981; Kojima et al., 1984). In some cases high concentrations of gelatin decreased aggregation (Wakiyama et al., 1982).

Solvent evaporation. Solvent evaporation may be accomplished by stirring the emulsion in an apparatus where the surface is exposed to air. Forced air or nitrogen may be used to promote a more rapid evaporation rate. Heat and reduced pressure may also be used but should be controlled at such a rate that microcapsules with a smooth surface are obtained if desired. Heat and low pressure may cause foaming of the emulsion system which should be avoided, particularly at the early stages of phase separation (Fong, 1988).

Stirring rate. After the emulsion containing the particles with appropriate size has been formed and before evaporation has begun, the stirring rate should be such that there is minimum aggregation of the droplets until the microspheres are hardened. The main factors that control the particle size are speed, equipment and the concentration of the polymer in the dispersed phase and the concentration of the hydrophilic polymer or surfactant in the aqueous phase. Particle size tends to decrease and the size range is narrowed as the mixing speed increases (Benita et al., 1984). Stirring speeds of 800–1600 r.p.m. have been used when either gelatin or polyvinyl alcohol have been used as the emulsifier. At 900 r.p.m., small holes were observed in the microspheres, while at 500 r.p.m. they were absent (Nozawa and Higashide, 1978).

Reactor design. The use of baffles minimizes the vortex which can lead to microsphere aggregation and, in addition, droplet breakup occurs with the result that the average size of the microcapsules is decreased and the microsphere yield is increased (Bodmeier and McGinity, 1987c).

Evaporation or removal with an immiscible organic liquid, a non-solvent for the polymer

Core. The core should have a minimal solubility in the immiscible organic liquid in order that most of the core is encapsulated. Tartrazine has been encapsulated with cellulose acetate trimellitate (Sanghvi and Nairn, 1991) when evaporation of the polymer solvent is not permitted. When evaporation is allowed to proceed, several pharmaceuticals have been encapsulated: tetracycline, loperamide, metoclopramide, hydrochloride and also biological material (Maharaj *et al.*, 1984) and drug–resin complexes (Sprockel and Price, 1990).

Wall polymers. The polymers used in this process should not dissolve in the suspending medium, for example, ethylcellulose dissolved in acetone was dispersed in a non-volatile hydrocarbon liquid (Dispersol 81515) (Yoshida, 1972). Another example is cellulose acetate phthalate dissolved in a mixture of acetone and ethanol, 95%, and then added to mineral oil (Beyger and Nairn, 1986). The size of the microcapsules increased from an average diameter of 140 μm to 295 μm when the concentration of the polymer, Eudragit RS was increased two and a half times. The reasons for this increase in size are attributed to the increase in viscosity of the dispersed phase as a result of higher polymer concentration and an increase of polymer inside the droplets affecting a larger volume (Pongpaibul *et al.*, 1984).

Polymer solvents. In order to avoid evaporation, the use of polymer solvent should be selected so that it has some solubility in the immiscible organic liquid, thus avoiding the use of temperature and the destruction of heat labile drugs. The removal of acetone, which has limited solubility in mineral oil, effects the phase separation of the polymer cellulose acetate trimellitate and subsequently microcapsules are formed (Sanghvi and Nairn, 1991, 1992).

Surfactants. The use of surfactants with low HLB values increases the region of the phase diagram where microcapsules were formed. The use of these surfactants tends to give a smooth surface on the microcapsules and at 3% concentration gives smaller microcapsules. Surfactants with a higher hydrophilic lipophilic balance value decrease the region on the triangular phase where microcapsules could be produced (Sanghvi and Nairn, 1991).

Immiscible organic liquid. Mineral oil is used extensively in this process of phase separation. It has been used in preparation of microcapsules with the following polymers: polymethyl methacrylate (Sprockel and Price, 1990), cellulose acetate phthalate (Beyger and Nairn, 1986), and ethylcellulose (Kaeser-Liard *et al.*, 1984).

THEORY AND MECHANISM

This section is concerned with some physical and chemical parameters, mechanisms and theories and/or experimental evidence which support the various concepts for coacervation-phase separation and deposition of the coacervate onto the core. As a result, this section is split into three parts:

1. A single wall-forming polymer soluble in water
2. Two wall-forming polymers soluble in water
3. A single wall-forming polymer soluble in an organic liquid

A single wall-forming polymer soluble in water

Phase diagrams

In order to prepare a satisfactory solution of a water-soluble polymer, it is necessary to disperse the polymer in water and allow it to become fully hydrated, perhaps using appropriate temperature conditions, addition of a small amount of non-solvent and/or mechanical means.

Coacervate-phase separation is induced by a number of techniques such as addition of a water miscible solvent, the non-solvent, which is not a solvent for the polymer, or a salt that binds a considerable amount of water thus removing it from the polymer. This process can best be illustrated by three component phase diagrams. The components usually are polymer, solvent and the agent which effects coacervation such as salt or ethanol. An example is given in Fig. 2 for gelatin, water and ethanol (Nixon *et al.*, 1966). It can be seen that a dilute solution of gelatin in water, say 10%, will form a single phase. As ethanol is added, the two-phase region is encountered and the polymer-rich phase, the coacervate, is produced which encapsulates the core material if present. The polymer-poor phase functions as the medium to suspend both the coacervate and core which is being encapsulated. At pH values distant from the isoionic point, flocculation occurs in the presence of ethanol because the gelatin is fully stretched and is unable to entrap the occlusion liquid. The flocculation region is generally not satisfactory for encapsulation. The concentrations in the various regions in this system were studied by using refractive index and specific gravity (Khalil *et al.*, 1968; Nixon *et al.*, 1966). Rigidization of the coat may

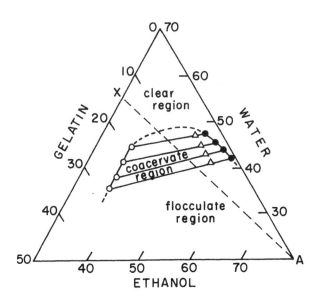

Fig. 2 The composition of coacervate and corresponding equilibrium liquid. ○ Coacervate; ●
equilibrium liquid; △ total mixture. Reproduced with permission from Nixon *et al.* (1966), *J.
Pharm. Pharmacol.* **18**, 409–416. The Royal Pharmaceutical Society of Great Britain, London.

be effected by temperature change, use of a cross-linking agent and/or the
use of appropriate non-solvent for the coat.

Phase diagrams are also useful for preparing nanoparticles which are in
the nanometre size range. For example, one method of preparing nanoparticles from gelatin or albumin is to desolvate the polymer with a salt which
is highly hydrated, thus causing the coacervate to form. Then the protein
is just resolvated with a small amount of water or isopropanol. In this procedure the phase diagrams are prepared using light scattering to measure the
onset of coacervation and thus the appropriate conditions to achieve
nanoencapsulation. The nanoparticles can be rigidized with a suitable cross-
linking agent such as glutaraldehyde (Oppenheim, 1986). Thus, phase
diagrams are useful for preparing nanoparticles which form in the precoacervation region, microcapsules or microspheres which separate in the coacervation region, and purification of the protein in the flocculation region
when excess salt is added to the system (Oppenheim, 1986).

Hydrogen ion concentration

Khalil *et al.* (1968) investigated the role of pH in the coacervation of gelatin. Since gelatin exists as a randomly coiled configuration in solution, the shape of these coils is influenced, by the ionization of the acidic and basic groups. A stretched configuration is predominant when the gelatin is mainly in the anionic or cationic form. A random coiled structure is favoured at the isoionic point as a result of inter- and intra-molecular attractive forces. The role of pH as it affects coacervation was explained by two factors which appear to influence coacervation of polyelectrolyte systems, namely inter- and intra-molecular attractive coulombic forces, and hydration. The authors postulate that the first of these effects favours phase separation and formation of floccules while the second promotes redispersion of the molecular species. A proper balance between these factors promotes the formation of a colloid-rich isotropic liquid phase which is the coacervate. At the isoionic point there is a balance between the attractive forces of the oppositely charged sites and the hydration effect. The authors found that at the isoionic point, a coacervate was readily obtained when ethanol was added. As pH values move away from the isoionic point, attractive forces decrease and hydration of the gelatin increases; both of these tend to prevent coacervation. At pH values considerably different from the isoionic point, flocculation occurs upon addition of ethanol as the molecule is stretched and cannot entrap sufficient water. At intermediate pH values, a viscous gel is formed. There is more flexibility in the gelatin chain but insufficient water is entrapped to form a coacervate.

When sodium sulfate is used as the coacervating agent, the ions in solution shield the charges on the gelatin material and thus the forces of repulsion are modified by the added salt. At pH values on the acidic side of the isoionic point of gelatin the sulfate ions are associated with the positively charged groups and coacervation proceeds satisfactorily.

Microcapsule formation

Madan (1978) has suggested that deposition of the polymer onto a core may take place as a result of:

1. Molecular interaction between the colloidal macromolecular particles,
2. Coacervate droplets may coalesce about the core particles,
3. Single droplets may encompass one, or a group of core particles.

An examination of the surface characteristics of microcapsules of gelatin, prepared by coacervation using ethanol or Na_2SO_4, showed that the dried microcapsules had no cracks or fissures. The smoother surfaces of the microcapsules produced using ethanol showed marked surface folding, attributed to vacuole formation in the coacervate droplet, which increased with time, allowing for the formation of the coacervate coat. As a result of these observations, Nixon and Matthews (1976) proposed that microcapsules prepared by coacervation resulted from the merger of several smaller microcapsules.

Encapsulation of liquids

Several microencapsulation procedures involve the encapsulation of immiscible liquids which may, or may not, contain a drug dissolved, dispersed or emulsified in the liquid. In order for the process of microencapsulation to proceed properly, the coacervate must engulf the liquid drop and then be hardened. In this system there are three immiscible phases present – the liquid core, the coacervate and the polymer-poor phase. Torza and Mason (1970), both theoretically and experimentally, investigated the interfacial phenomenon of systems that contain three liquids which are immiscible and indicated the spreading coefficient necessary for a coacervate droplet (liquid 3) to surround the core liquid (liquid 1) when both are in an immiscible continuous phase, the polymer-poor phase (liquid 2). The three spreading coefficients for the three-phase system are:

$$S_1 = \sigma_{23} - (\sigma_{12} + \sigma_{13})$$
$$S_2 = \sigma_{13} - (\sigma_{12} + \sigma_{23})$$
$$S_3 = \sigma_{12} - (\sigma_{13} + \sigma_{23})$$

In order to consider the process it is assumed that $\sigma_{12} > \sigma_{23}$ thus $S_1 < 0$, and thus only three possible sets of values of S exist. These correspond to complete engulfing, partial engulfing and non-engulfing of the liquid core. Complete engulfing occurs under the condition that $S_1 < 0$, $S_2 < 0$, $S_3 > 0$. As the interfacial tension between the polymer-rich phase, that is the coacervate, and the polymer-poor phase is low and if the interfacial tension between the core and the continuous medium is higher than between the core and the coacervate-rich phase, then the above requirements are satisfied and engulfing will occur. Torza and Mason (1970) conducted a number of experiments using different liquids and found that of 20 systems studied, only five did not correspond with the theory. The method of engulfing was determined with a high-speed movie camera and was shown to involve two competitive processes – spreading and penetration. This theory may apply to any coacervate system involving liquids.

In a study of the microencapsulation of oil droplets with or without a drug, clofibrate or chlormethiazole, using gelatin with an isoelectric point of 4.85 in the presence and absence of a surface-active agent, Siddiqui and Taylor (1983) were able to relate the ionic charge on the coacervate, which was negative, and the spreading coefficient of the liquid substrate. The surface active agents cetrimide, sodium lauryl sulfate or a double salt hexadecyltrimethylammonium lauryl sulfate were used in the process. Spreading coefficients calculated from interfacial tension values indicate that the coacervate should spread more easily in the presence of the double salt, and less so in the presence of either of the other two surfactants. It was noted that conditions for measuring the spreading coefficient and microencapsulation were not identical. Measurements of the charge size on the dispersed particles showed that the oil droplets and coacervate droplets can be expected to have opposite charges, except in the presence of sodium lauryl sulfate. Microencapsulation was satisfactory with both the double salt and cetrimide, but not with sodium lauryl sulfate. The authors suggest that some cetrimide may enter the coacervate phase and subsequently microcapsules tend to agglomerate. Addition of sodium lauryl sulfate at this stage tended to prevent agglomeration. The authors suggest that the use of the double salt enhances the attachment of the coacervate to the oil droplet surface. In fact, this double salt produces the smoothest microcapsule and permits a slower release of the drug from the oil droplets.

Adsorption studies

By electrophoresis and adsorption studies of gelatin and various core materials using six different kinds of coacervating agents, Okada *et al.* (1985a) were able to show that suitable encapsulation by gelatin is affected by the affinity between the core material and the coacervate phase. If a large amount of gelatin is adsorbed prior to coacervation, then encapsulation is successful.

In a subsequent paper Okada *et al.* (1985b) showed that carboquone could not be encapsulated with gelatin unless methanol or sodium sulfate solution was used as the coacervate inducing agent. If, however, the drug is recrystallized from a solution of an ionic polymer and if the pH of the solution is appropriate, then the drug can be encapsulated using many coacervating inducing agents. It was shown that the electrostatic attraction found between the gelatin and the polymer attached to the drug has an important role.

An incompatible or non-wall-forming polymer

Instead of using alcohols of low molecular weight to cause coacervation, Jizomoto (1985) used polyethylene glycol or polyethylene oxide. Both these polymers have the same general structural formula, but the former name is usually used for polymers with molecular weights $\leqslant 20\ 000$ and the latter for those greater than tens of thousands. Polyethylene oxide or polyethylene glycol was added to aqueous gelatin 2.2% with stirring at 40°C at various pH values to influence phase separation. The phase diagram shows that the addition of a small quantity of polyethylene oxide or polyethylene glycol causes phase separation over a wider pH range and this phenomenon is largely dependent on molecular weight. A plot of the log of minimum concentration of polyethylene glycol or polyethylene oxide required to effect phase separation against the log of molecular weight at pH 8.7 gives a straight line. Elemental analysis of the coacervate is identical with that of gelatin; thus the phase separation induced by the addition of polyethylene oxide or polyethylene glycol is caused by an incompatibility. After reviewing the theory proposed by Bailey and Callard (1959) which suggests that water molecules are orientated with respect to polymer chain and the postulation of Kagemoto et al. (1967) that the parameter ascribed to the interaction between the oxygen atoms of the ether bonding in polyethylene oxide chain and the water molecules is proportional to a function of the molecular weight, Jizomoto suggests that the effect of polyethylene oxide or polyethylene glycol on phase separation in relation to molecular weight cannot be explained by dehydration. Blow and coworkers (1978) suggested that polyethylene oxide causes a decrease in 'free water' but in the present experiments, polyethylene oxide concentration to bring 'free water' to zero was not appreciably influenced by molecular weight. Therefore, the dependence of phase separation on molecular weight cannot be explained by the concept 'free water'. The author used a simplified equation of the chemical potential expressed in terms of molalities of two polymers, the solvent and the exclusion volumes as described by Edomond and Ogston (1968). Calculations were made to estimate the minimum concentration of the polymers required to cause phase separation. The author concludes that the excluded volume should make the main contribution to the induction of phase separation.

Two wall-forming polymers soluble in water

Effect of ionic charge

Complex coacervation may be brought about in water by the combination of two polymers, one with a positive charge and the other with a negative charge. The most common polymers used are gelatin, which is dissolved in water and the pH is adjusted so that it is below the isoelectric point, and acacia which is negatively charged because of the ionization of its carboxyl groups. The combination leads to a coacervate, which is polymer rich, and the other phase which is polymer poor (Luzzi, 1976). The interaction between the two polymers is also influenced by temperature and the presence of salts (Madan, 1978).

Theory

In a series of papers Nakajima and Sato (1972) reported upon the phase relationships and theory of complex coacervation of the sulfated polyvinyl alcohol–aminoacetalysed polyvinyl alcohol system. Phase relationships were examined for the polymer salt, water and sodium bromide. The experimental results were interpreted by the use of a theoretical equation for the free energy of mixing by taking into account the entropy and enthalpy contributions ascribed to a non-ionic polymer solution and the electrostatic free energy expression as derived by Voorn (1956). In two subsequent papers, Sato and Nakajima (1974a,b) investigated the effects of chain length of the polyelectrolytes, the thermodynamic interaction between the polymer and water and the number of charges of polyelectrolyte chain on the complex coacervate on the basis of a free energy equation. Conditions for the formation of coacervate droplets as a function of charge density and polymer concentration were also discussed.

Burgess and Carless (1984) investigated the electrophoretic mobility profile of polyions and showed that these profiles can be used to determine if complex coacervation will occur between two polyions. Furthermore, they showed that the pH range of coacervate, the pH of optimum coacervation and the salt tolerance of the system can be predicted. They also showed that the maximum coacervate volume occurred at the electrical equivalence pH, that is when the charges on the two polyions are equal and opposite.

A practical analysis of complex coacervate systems has been published by Burgess (1990) who reviewed several theories which are now briefly described. Overbeek and Voorn (1957) postulated that the coacervation which takes place between gelatin and acacia is a competition between ionic attractive forces, which tend to bring the polyions together, and entropy effects, which promote the dispersion. The coacervate phase binds water between the loops of the polymer chains. The water in the coacervate contributes to the entropy and permits a number of arrangements of polymers. As a result the coacervate is fluid. Another theory which attempts to explain complex coacervation is the 'dilute phase aggregate model', developed by Veis and Aranyi (1960) to take into account the formation of complex coacervation when the product of the charge density and the molecular weight is low. The model postulates that complex coacervation occurs in two steps, as oppositely charged gelatins fuse, aggregate, and then rearrange to form the coacervation phase. The rearrangement occurs slowly and is formed by the gain in configuration entropy. Several differences between the two theories are described by Burgess (1990). Burgess and Carless (1985, 1986b) confirmed the two-step process and detected the presence of small aggregates by light scattering. Tainaka (1979, 1980) modified the Veis and Aranyi theory to indicate that the aggregate pairs in the dilute phase did not have specific charge pairing. Again, the dilute phase aggregates condense to form the coacervate, but the aggregates are present in both the dilute and coacervate phases. The aggregates overlap with each other in the coacervate phase and, as a result, there is a gain in electrostatic energy due to the increase in ion density in the overlapped domain. High molecular weights and highly charged densities of the polyions enhance the attractive forces effecting phase separation. The Tainaka theory explains the suppression of coacervates at high polymer concentration as stabilization of aggregate structures at high concentration. Burgess (1990) concluded that, while the Tainaka model is not all-inclusive, as it does not explain the reduction of coacervation at low ionic strength, it is not as restricted as other theories and thus, at present, is the best general theory.

Polar bonding

A recent paper by Van Oss (1988–1989) presents a somewhat different classification of coacervation, complex coacervation and flocculation based on polar (hydrogen) bonding components of interfacial interactions. Van Oss (1988–1989) has reviewed the classification for coacervation (simple) and complex coacervation for the system of gelatin and acacia (Table 1). A theoretical analysis of cohesion and adhesion in terms of the Lifshitz–van der Waals, or apolar, components and Lewis acid–base, or polar, components of free energy between two different bodies 1,2, through a liquid 3, indicates interfacial (hydrophobic) attraction when $\Delta G_{132} < 0$ and interfacial ('hydration pressure' mediating) repulsions when $\Delta G_{132} > 0$. As a result of this theory, Van Oss provides a table which indicates the mechanisms and conditions for coacervation (simple) and complex coacervation. He indicates that coacervation (simple) takes place when polar and/or apolar repulsion between the two solutes, where one or both must be a polymer dissolved in the same solvent, results in phase separation. Complex coacervation takes place when there is electrostatic or polar attraction between two polymers of opposite charge (or of opposite signs of Lewis acid–base behaviour). Examples of coacervation (simple) due to polar interactions are negatively charged gelatin and gum arabic, agar and ethanol, polyvinyl alcohol and polyethylene glycol, the solvent in all cases being water.

Table 1 Comparison of coacervation (simple) and complex coacervation using a mixture of gelatin and acacia given by Bungenberg de Jong, adapted from Van Oss (1988–89)

Property	Coacervation (simple)	Complex coacervation
pH	>Isoelectric point for gelatin	<Isoelectric point for gelatin
Concentration of original solutions	Occurs with concentrated solutions	Occurs with dilute solution
Indifferent salts	Tend to promote coacervation Place in lyotropic series is important	Tend to suppress coacervation Place in lyotropic series is minor Valency is important
DC field	Drops show no disintegration	Drops show disintegration
Composition of liquid layer	Each layer contains essentially one species	The coacervate layer is rich in the colloid which is a ratio of about one to one
Principal condition	Water deficit in the system	Different charge between the two species

Coacervation (simple) due to apolar interactions includes cellulose acetate and ethanol dissolved in chloroform, polyisobutylene and polystyrene in benzene. Complex coacervation always takes place in water and some examples resulting from electrostatic interaction are positively charged gelatin and negatively charged gum arabic, and positively charged gelatin and nucleic acid. Examples of polar (Lewis acid-base) interaction include polyacrylic acid and polyvinyl methylether.

Borue and Erukhimovich (1990) developed a microscopic statistical theory of symmetrical polyelectrolyte complexes. The complex was shown to form a polymer globule and the equilibrium density, the width of the surface layer and the surface tension were calculated as a function of salt concentration. Complex coacervation is considered as a precipitation of polymer globules due to a minimization of surface energy. The theory is based on the Lifshitz–Grosberg theory of polymer globules and the authors' previous work concerning the equation of state of polyelectrolyte solutions.

Particle size

In a study of the encapsulation of hydrophobic compounds such as stearyl alcohol by complex coacervation with gelatin–acacia, Madan *et al.* (1972) found that only particles below 250 μm diameter could be encapsulated. It was proposed that the mechanism for encapsulation in this system was either a single coacervate droplet which encompasses a group of immiscible nuclei or individual coacervate droplets adsorbed to, or coalesced around the particles. Photomicrographs of 163 μm particles indicate that encapsulation takes place by aggregation or coalescence of several droplets (with diameters usually under 40 μm) to surround the core stearyl alcohol particles. Larger particles were incompletely covered. The authors suggest that the affinity of the coacervate droplets for the core material is not great. They suggest that the velocity difference between the core particles and the coacervate droplets, as the mixture is stirred, tends to prevent the aggregation of droplets around the core particles. In addition, the probability that a sufficient number of coacervate droplets will aggregate and coalesce to surround a core particle decreases as the particle size increases. Experimental studies showed that larger particles could be encapsulated if the concentration of the coacervate was increased. In order to improve the encapsulation process, the stearyl alcohol was melted in an acacia solution and then congealed. The acacia is adsorbed more strongly to liquid stearyl alcohol than to the solid form. The adsorbed acacia then reacts directly with the gelatin in the encapsulation process to form the microcapsules.

Thermodynamics

Veis (1975) described the thermodynamics of phase separation in a mixture of oppositely charged polyelectrolytes. He indicated that homogeneous solutions will be formed as long as a plot of ΔG_M, the free energy of mixing of a solute and solvent, versus ϕ_2, the volume fraction of the polymer, has a single minimum. However, if X_{12}, the interaction parameter, which is proportional to the interaction energy per mole of solvent molecules, is sufficiently large and positive two minima will be present in the plot and any mixture prepared between these two will separate into two phases. Mathematical analysis shows that for polymers of moderate size, phase separation will occur at low solute volume fraction if X_{12} is slightly greater than 0.5.

The thermodynamics of mixing of two dissimilar polymers in a single solvent are also discussed. The author discusses two cases. In the first case the polymeric ions have a very high charge density and phase separation occurs to give essentially solvated coprecipitates in equilibrium with an extremely dilute phase. These precipitates are the basis of certain membranes. The other case is that in which the polyions are of a moderate charge density and phase separation is driven by the more favourable electrostatic interaction in the concentrated phase. In this example, both phases contain both ionic polymers, as is the case for the gelatin–acacia interaction.

Based on two experimental findings, namely charge equivalence in the coacervate phase and molecular weight pairing in the coacervate phase, the author suggests the possibility of two mechanisms for the formation of the coacervate based on an unfavourable entropy change ΔGentropic > 0 and a favourable electrostatic free energy change ΔG electrostatic < 0 for the reaction:

$$P^+OH^- + H^+Q^- \rightarrow [PQ]Agg + H_2O$$

where P^+ is the cationic polymer, Q^- is the anionic polymer and $[PQ]Agg$ is the aggregate.

The aggregation may take two forms: the two molecules with the centres of gravity overlapped, or two molecules with explicit ladder-like charge pairing. The author argues in favour of the ladder type formation, based on the molecular weight pairing and the suppression of coacervation observed in most polydispersed mixtures. This new aggregate, PQ, behaves as a new component which should obey the basic Florey–Huggins polymer binding mixture phase separation rule.

Surfactant effects

The influence of cationic, anionic and non-ionic surfactants on complex coacervate volume and droplet size has been researched by Duquemin and Nixon (1985). The coacervate was prepared by dissolving the surfactant in the acacia solution and then adding an equal quantity of gelatin solution at the optimum pH of coacervation, 4.35. It was found that the per cent coacervate weight decreased with increasing concentration of sodium lauryl sulfate. At high surfactant concentrations, 0.20 and 0.35%, and at a low colloid concentration, 1%, formation of the coacervate was prevented. It was postulated that the additional ions from the surfactant prevented or restricted electrostatic attraction between the gelatin and acacia polyions. The effect of increasing concentration of cetrimide on the per cent coacervate by weight is not so clear and depends on the concentration of the colloid. At low concentration of the surfactant, there is a slight increase in weight and this has been attributed to an increase in water content of the coacervate. At a high colloid concentration, 4 and 5%, and high surfactant concentration, 0.075%, there is an appreciable decrease in the concentration of the coacervate. The authors suggest that this is due to the suppression of coacervation because the process is less energetically favourable. The effect of polysorbate 20 on coacervate weight is similar to that produced by cetrimide. It is suggested that steric hindrance of the large surfactant molecules suppresses coacervation.

A single wall-forming polymer soluble in an organic liquid

Solubility effects

Use of a non-solvent. After the polymer is dissolved in an appropriate organic solvent, phase separation may be induced in a number of ways. For example a second organic solvent which is miscible with the solvent for the polymer, but at the same time is a non-solvent for the polymer may be added. The solubility of the polymer is now decreased and separates under appropriate conditions as a polymer-rich phase which is also able to surround the core. The polymer-rich phase may be hardened by further treatment with the miscible non-solvent (Luzzi, 1976).

Relation between polymer composition, solvent and non-solvent. The mechanism of coacervate formation in a non-aqueous system has been investigated by Ruiz *et al.* (1989). Microcapsules of poly (DL-lactic acid-co-glycolic acid) were prepared by dissolving the polymer in methylene chloride and then adding various quantities of silicone oil to effect phase separation. The phase separation phenomenon was observed by taking photomicrographs after increasing quantities of the incompatible polymer were added. At the first step when the amount of phase inducer is low (1–5%), a pseudo-emulsion of the silicone liquid is formed. During the second step when more silicone oil is added, the beginning of the phase separation appears. The coacervate droplets appear to be unstable and merge together then break apart. In the third step the added quantity of silicone oil is sufficient to permit a stable dispersion of polymer coacervate droplets; this step is called the stability window. Finally, the fourth step occurs after further addition of silicone oil, which causes extensive aggregation of the coacervate droplets.

Four polymers with different compositions of lactic acid (LA) and glycolic acid (GA) were studied. The polymers with the highest percentage of lactic acid had the largest stability window, and required the largest amounts of silicone oil to reach that region. The polymer with the lowest content of lactic acid, namely 50% LA and 50% GA, is the least hydrophobic among the polymers studied and did not easily dissolve in the solvent and required only small amounts of silicone oil to effect phase separation, accounting for the small stability window. Silicone oil with a low viscosity did not yield a stability window with four polymers tested; however, silicone oils with higher viscosity, up to 12 500, increase the size of the stability window for all polymers. The stability window was increased by increasing the solubility of the poly(DL-lactic acid-co-glycolic acid) polymer in the solvent by the addition of methanol. As a result, more silicone oil was needed to reach the stability window, and thus to induce the appearance of the polymer droplets. The width of the stability window can be altered by changing the viscosity of the silicone oil and modifying the solvent for the polymer by adding a suitable percentage of a better solvent.

Shively and McNickle (1991) considered the effect of various solvent compositions on the coacervation process. Ternary phase diagrams were prepared using a biodegradable block copolymer prepared from tartaric acid and 1,10-decanediol, and ethanol and water, with or without NaCl. Microcapsules of kaolin or hydrocortisone-21-acetate were prepared by adding the core to an ethanol solution of the polymer and then titrating with the aqueous non-solvent. The microcapsules were then filtered and dried. At high polymer concentrations in the non-plait region, minimal or no solvent interaction occurred and the polymer was in the coiled configuration. In the plait region at low polymer concentration, the ionic strength of the non-solvent showed an effect on the coacervate, the adhesive forces were greater than the cohesive forces, and the polymer adopted a more linear configuration. Surface tension measurements, when the solvent composition was 30% water and 70% alcohol (non-plait) or 50% alcohol and 50% water (plait), showed that the area per polymer molecule decreased in post-coacervation compared with the precoacervation region. These results agree with the theory that the coacervation results in the reduction of the surface free energy of a system through a reduction of the molecular surface area. Analysis of the surface tension versus polymer composition graphs shows that coacervate phases resulting from 30% water, 70% ethanol compared with 50% water, 50% ethanol were very different suggesting differences in molecular configuration and interaction properties. Thus, the authors speculate that microcapsules made with different coacervation conditions would have different properties, such as diffusion or morphology. It was found that microcapsules produced with non-plait conditions had considerably slower rates of release of the drug and had rough and irregular surfaces compared with microcapsules prepared with plait conditions.

Solubility parameters. Robinson (1989) determined the solubility of ethylcellulose Type N10 in 122 solvents qualitatively and in 36 solvents quantitatively. The contribution of dispersive, polar and hydrogen-bonding intermolecular forces was determined and plotted on two-dimensional and triangular solubility graphs. The influence of dipole–dipole interactions on the solubility of ethylcellulose was shown by plotting the fractional polarity of the solvent against the solubility parameter. The diagram shows that ethylcellulose is soluble over a range of polarity from 0 to 0.75, but it is not soluble in solvents with either a low or high solubility parameter. In order to show the effect of the relative fractional contributions of the hydrogen bonding, polar and dispersion components, a triangular solubility diagram was prepared. The solvents were classified on their hydrogen bonding ability: weak, medium, and strong. The three areas of solubility overlap and they define a region which determines the intermolecular forces appropriate to dissolve ethylcellulose. Ethylcellulose occupies a central position within the defined solubility regions. The triangular diagram is useful for determining good solvents and non-solvents for microencapsulation purposes. Coacervation was observed after cooling a solution of ethylcellulose in a poor solvent which had a solubility parameter near, or just outside, the solubility region for the polymer. Gelation occurs after cooling with liquids usually considered non-solvents for the polymer and their solubility parameters are well outside the solubility region and have higher interaction parameters. Flocculation was observed after cooling solutions of the polymer in polar solvents where large values of the interaction parameters occur.

The selection of appropriate solvents and non-solvents may be ascertained through the use of solubility parameters. In a study by Moldenhauer and Nairn (1992), it was shown that microcapsules could be prepared by phase separation using a number of solvent–non-solvent pairs. The solubility parameter map was prepared using a number of solvents, both singly and in mixtures, to provide regions where the polymer ethylcellulose was

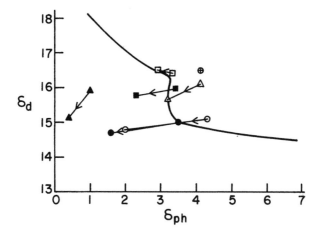

Fig. 3 Solubility parameter map for ethylcellulose showing the solubility border and the initial and final microencapsulation solubility parameters: □ using ethyl acetate and cyclohexane; ⊕ using methyl ethyl ketone and cyclohexane; ▲ using toluene and light liquid paraffin; ■ using ethyl acetate, cyclohexane and light liquid paraffin; △ using methyl ethyl ketone, cyclohexane and light liquid paraffin; ● using ethyl acetate and light liquid paraffin; and ○ using methyl ethyl ketone and light liquid paraffin. Reproduced with permission from Moldenhauer and Nairn (1992), *J. Controlled Release* **22**, 205–218. Elsevier Science Publishers BV, The Netherlands.

soluble (that is, gave clear solutions at definite concentrations) and regions where the polymer was insoluble (that is, where clear solutions were not obtained). This information was then used to prepare microcapsules of theophylline ion-exchange resin beads coated with ethylcellulose using a number of solvents. Partial evaporation of the solvent in a mixture leads to a change in solubility parameters effecting phase separation. These authors experimentally corroborated Robinson's (1989) studies that microencapsulation systems should be near the limit of ethylcellulose solubility where coacervation will occur and showed that microcapsules could be prepared by controlled evaporation of a solvent–non-solvent pair for ethylcellulose. The total amount evaporated was the same in all experiments. As evaporation of the solvent and non-solvent took place, the solubility parameter of the mixture generally changed, owing to the different vapour pressures, into a poor solvent for the polymer. The composition of the solvent pair, both before and after evaporation, was related to the phase diagrams and the solubility parameter map. Well-formed microcapsules were prepared from solvent–non-solvent pairs whose solubility parameters changed during evaporation from the soluble region to just at the other side or at the edge of the solubility region on the solubility parameter map. Even though different solvent mixtures were used, ethyl acetate or methyl ethyl ketone as solvents and non-solvents cyclohexane and light liquid paraffin, the solubility parameters were similar and evaporation produced microcapsules with similar characteristics (see Fig. 3).

Experiments conducted with solvents that had a poor solubility parameter for ethylcellulose yielded either no coat or a coat of poor quality. Evaporation of solvent pairs which had similar vapour pressures and which were in the solution region of the solubility parameter map did not change their solubility parameter during evaporation, and microencapsulation did not take place. Solubility parameter maps provide information about a number of solvent–non-solvent pairs whereas a phase diagram provides information about only one solvent–non-solvent pair.

Wall formation

The mechanism of wall film formation of ethylcellulose onto magnesium aluminium hydroxide hydrate was investigated by Kasai and Koishi (1977). Four different experiments were carried out in order to investigate the phenomenon of microencapsulation. With increasing amounts of ethylcellulose, in dichloromethane, added to the core and also addition of water, it was shown that the surface properties of the core changed from hydrophilic to hydrophobic, likely as a result of adsorption of the ethylcellulose onto the surface of the core. Photographs of ethylcellulose coacervate drops formed by the addition of increasing volumes of *n*-hexane to the ethylcellulose solution show an increase in size of the coacervate drops which corresponds to an increase in the weight of ethylcellulose in the coacervate. The authors indicate that the smaller-sized coacervate droplets are likely to be suitable for microencapsulation, and the larger-size coacervate droplets are not. The authors also relate the surface structure of the microcapsules to the weight of ethylcellulose coacervate as a result of the addition of non-solvent.

The authors then suggest possible cross-sectional models for the deposition of ethylcellulose coacervate drops on the core and in the final state of the walls as shown in Fig. 4. It is noted in Fig. 4 that there is compression of the ethylcellulose on the core material and this is supported by the fact that despite an increase in percentage ethylcellulose concentration in the microcapsules from 27 to 46%, the wall thickness range is almost constant within the range of 15–17.5 μm. In conclusion, the authors postulate a model for deposition of ethylcellulose coacervate based on the amount of coacervate on the core, as a result of increasing amounts of non-solvent added, scanning electron microscopy of the microcapsule surface, photographs of the coacervate droplets and the region of constant wall thickness.

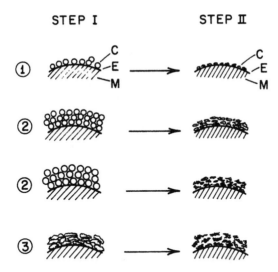

STEP I   STEP II

Fig. 4 Possible cross-sectional models for the deposition of ethylcellulose coacervate drops on the core material and the final state of microcapsule walls. Key: step I: coacervate drops deposited at first; step II: final walls; ① ② and ③: different stages, C, E, and M, coacervate drops, ethylcellulose adsorbed initially on the core material, and the core material. Reproduced with permission from Kasai and Koishi (1977), *Chem. Pharm. Bull.* **25**(2), 314–320. Pharamaceutical Society of Japan.

Incompatible or non-wall-forming polymers

A number of papers have been written about the effect of polyisobutylene on the coacervation of ethylcellulose and the formation of microcapsules. In an early paper Donbrow and Benita (1977) describe ethylcellulose coacervation by dissolving the polymer in cyclohexane and slowly cooling with controlled agitation. Phase separation occurs over 24 h to yield a lower phase of coacervate droplets and a clear upper layer containing polyisobutylene. The authors noted the non-linear increase in volume of the coacervate with polyisobutylene concentration. At the same time, a decrease in the particle size of the coacervate droplets was related to the increase in phase coacervation volume effected by the change in polyisobutylene concentration. They attributed the rise in phase coacervation volume with polyisobutylene concentration to an increase in the volume of adsorbed solvated polyisobutylene. The adsorbed layer minimizes agglomeration of the droplets rather than the mixing of their solvated polyisobutylene layers. They suggest that polyisobutylene acts as a protective colloid in the coacervation process and prevents the formation of large aggregates of ethylcellulose. A free-flowing powder was obtained on drying when polyisobutylene was employed; however, in the absence of polyisobutylene an aggregate mass was produced. Benita and Donbrow (1980) extended their research on the role of polyisobutylene and its effect on coacervation. Microanalysis indicated that polyisobutylene was not coprecipitated with the washed ethylcellulose coacervate droplets and thus functions as a stabilizer by adsorption. The increase in phase coacervation volume with increasing polyisobutylene concentration was explained by a decrease in sedimentation rate as a result of combined effects of the smaller size of the droplet and the higher viscosity of the medium. The final phase coacervation volume is determined less by the ethylcellulose close-packed volume than by the repulsion forces between the stabilized droplets. During the cooling process in order to solidify the coacervate drops, the adsorbed layer of polyisobutylene increases the surface viscosity and it is expected that the rate of surface nucleation of the ethylcellulose decreases, thus explaining the formation of the smooth surface, characteristic of a structure of amorphous nature. The process would be promoted by an increase in adsorption of polyisobutylene and surface viscosity as the temperature falls. The authors concluded that polyisobutylene, a linear polymer, acts by forming a high-energy barrier as a result of adsorption of anchor groups on the surface of the droplet; the rest of the polyisobutylene molecule, bound by either looped segments or segments, is directed toward the outside of the coacervate droplet. This arrangement provides steric stabilization as a result of repulsion of solvated polymer chains.

The mechanism of aggregation prevention by polyisobutylene was studied using Eudragit RS or RL as the wall polymer, tetrahydrofuran as the solvent and cyclohexane as the non-solvent (Donbrow *et al.*, 1990). Phase diagrams, phase volume ratios of the system and photomicrographs of various stages of microencapsulation were presented. The presence of polyisobutylene permits the formation of two liquid phases, which is an unstable emulsion and a significant volume fraction is occupied by the wall polymer phase. During addition of the non-solvent cyclohexane, the solvent tetrahydrofuran is removed from the wall polymer phase and its volume fraction decreases. The dispersed concentrated wall polymer phase remains fluid during deposition onto the core surface and then gelling occurs and the system is then composed of two liquids and one gel. If polyisobutylene is not present a viscous coacervate rapidly separates, which is adhesive, during the slower desolvation stage as the composition of the solvent changes, and in this case liquid and gel are formed. The authors suggest that secondary dispersion phase phenomena are more readily controlled in the two liquid dispersion compared with liquid gel dispersion and the polyisobutylene permits steric stability. The effect of polyisobutylene molecular weight was also investigated. Polyisobutylene (mol. wt 50 000) did not permit the formation of microcapsules, but yielded matricized core particles. This was attributed to the formation of a low volume wall-polymer phase of high viscosity, even prior to non-solvent addition, which was not able to provide an appropriate coating similar to the condition of the gel formed in the absence of polyisobutylene. Polyisobutylene of higher molecular weight gave two incompatible fluid phases which progress through the appropriate changes on addition of cyclohexane to give microcapsules. Polyisobutylene solutions at a constant viscosity but using higher molecular weight polymers, that is, using smaller concentrations of polyisobutylene, gave smaller phase volumes at initial condition and smaller droplet diameters.

The influence of coacervation inducing agents, namely butyl rubber, polyethylene and polyisobutylene, was studied by Samejima et al. (1982). The core was ascorbic acid and the coat of ethylcellulose was deposited on the core drug as a result of temperature reduction from a solution of the polymer in cyclohexane. Scanning electron microscopy of the resulting microcapsule showed that the surface of ascorbic acid was poorly covered when butyl rubber was used as the coacervation inducing agent, a smooth coat with small holes was produced when polyethylene was used and a smooth surface with few holes was obtained with polyisobutylene. The wall thickness was in the range from 0.75 to 3.72 μm and was in the order butyl rubber < none < polyethylene < polyisobutylene. The authors suggested that the factors of smoothness and wall thickness influenced the dissolution rate in the order, butyl rubber > none > polyethylene > polyisobutylene. In conclusion, polyisobutylene is adsorbed on the coacervate wall which is on the surface of the crystal and functions as a stabilizer, preventing the agglomeration of single microcapsules into aggregates.

Viscosity and surface tension effects on microsphere size

The influence of viscosity and surface tension on the particle size of microspheres prepared by emulsification was investigated by Sanghvi and Nairn (1992). The microspheres were prepared by dissolving different amounts of polymer, cellulose acetate trimellitate, in solutions of acetone and ethanol and adding this solution to the external phase composed of mixtures of light and heavy mineral oil. The viscosities of the two phases were determined both before and after mixing and the interfacial tension between the two phases was also determined. The interfacial tension ranged up to 7 dynes cm^{-1} but did not affect the particle size appreciably. It was found that as the viscosity ratio of the internal phase to the external phase both before and after mixing increased, the particle size of the microspheres slowly increased from about 100 μm to about 200 μm until a minimum viscosity ratio of approximately 10 before mixing and approximately 1000 after mixing was achieved; subsequently there was a very rapid increase in the size to about 700 μm. The data were related to the theory of drop deformation as described by Becher (1965) and the ease with which particles coalesce (Gopal, 1968).

In a subsequent paper Sanghvi and Nairn (1993) were able to control the particle size of the cellulose acetate trimellitate microspheres by adjusting the ratio of the polymer to solvent concentration and by adjusting the internal phase volume fraction. The amount of polymer has a direct influence on the viscosity of the internal phase, and hence the viscosity ratio of the internal to external phase as described above. The phase volume ratio affects the particle size of the microsphere as it changes the probability of two droplets colliding and forming a larger droplet.

COACERVATION-PHASE SEPARATION USING A SINGLE WALL-FORMING POLYMER SOLUBLE IN WATER

This section and the next two are generally grouped according to the expected number of polymers in the wall and their solubility. Within each of these sections the polymers are arranged alphabetically and within these sections chronologically grouped according to the authors.

Acacia

Acacia (gum arabic) has been used to encapsulate oil drops such as lemon and polybutadiene. It has been reported that the coacervation process was more satisfactory if acacia was treated with acidic and basic ion-exchange resins to produce a salt-free form (Schnoering and Schoen, 1970).

Albumin

Research by Ishizaka et al. (1981) describes the preparation of egg albumin microcapsules and microspheres by dispersing a solution of the albumin, containing a core material, in isooctane containing sorbitan trioleate and then heating to about 80°C to cause denaturation of the albumin. It was found that the size distribution of the albumin microspheres was strongly affected by the surface-active agent concentration, mechanical agitation and albumin concentration.

Serum albumin microcapsules have been prepared by dispersing poly(acrylonitrile) beads of specific sieve fractions into a bovine serum albumin solution adjusted to pH 5.0 (Ishizaka et al., 1985). Isopropyl alcohol was added to the suspension to form the coacervate drops at 25°C. Vigorous stirring was employed to prevent the formation of multinuclear microcapsules. Subsequently, the suspension was heated to 70°C to harden the microcapsule wall. Coacervation of serum albumin was observed and a three component phase diagram was prepared. The optimum concentration for microcapsule formation was also provided on the phase diagram. Low concentrations of isopropyl alcohol were not appropriate for microcapsules, but 30% was found to be satisfactory. The characteristics of the product depended upon the total surface area of the core beads and the albumin concentration. At high surface areas of the core and low albumin concentrations, non-spherical and multinuclear microcapsules were obtained.

Several patents have been published by Ecanow and Ecanow (1983) describing the coacervate formed from albumin and lecithin for parenteral purposes. A composition containing albumin, urea, sodium chloride and lecithin solutes was prepared and stored at 4°C to give a coacervate which was then treated with cholesterol, $CaCl_2$ and KCl and the pH adjusted to 7.3 and made isotonic. After storage for 6 days, two phases are formed. It is suggested that it be used as synthetic whole blood.

In a subsequent patent, Ecanow (1988a) used human serum albumin for the solubilization and parenteral delivery of a drug dispersion. Butanol was added to a solution of egg lecithin and after shaking the middle phase was removed. To this phase, human serum albumin was added and dissolved. After storage in the cold, diazepam was added to the colloid-rich phase. The colloid-poor phase was then added and the mixture emulsified and the pH adjusted to 7.3–7.4.

Polymerized albumin and lecithin have been used to form a coacervate of erythromycin, and the dried particles used in oral products such as tablets, capsules and syrups (Ecanow, 1988b). Ecanow (1991) also described a coacervate containing egg albumin and egg lecithin which could be used for different routes of administration. In one example, bovine insulin was incorporated into the coacervate of egg albumin and then administered to rats. This resulted in a decrease in glucose blood levels whereas the unencapsulated insulin had no effect.

Simple coacervation with albumin has also been studied to prepare microcapsules of the core, namely sulfamethoxydiazine or acrylonitrile styrene copolymer resin beads. It was found that the time for 50% release increased from 6 min to 73 min when the drug was encapsulated with albumin (Ku and Kim, 1987).

Alginate

Sodium alginate has been occasionally used to prepare microcapsules. Salib *et al.* (1978) dispersed various drugs such as chloramphenicol or sulfadiazine in the sodium alginate solution and this mixture was added to an aqueous calcium chloride solution. Calcium alginate is formed and deposited around the drug particles. Loss of drug which was less than 16% occurred during the microencapsulation process.

Carboxymethylcellulose

Carboxymethylcellulose sodium has been used in the preparation of microcapsules by coacervation. In order to prepare microcapsules of indomethacin, the drug was dispersed in the aqueous polymer solution and this was added to an aluminium sulfate solution. Drug loss during the preparation of the microcapsules was minimal. The release of the drug followed an apparent zero process and there was a four- to eight-fold reduction in the release rate compared with the uncoated drug. The zero-order rate constant could be related to the coating ratio (Salib *et al.*, 1989).

Cellulose acetate phthalate

Cellulose acetate phthalate has been used to prepare microcapsules of phenacetin as described by Merkle and Speiser (1973). The polymer was dissolved in an aqueous solution containing a stoichiometrically equivalent quantity of Na_2HPO_4. The stirred solution was maintained at 60°C and the drug was added then the coacervating agent, a solution of sodium sulfate at 60°C. Subsequently, the solution was slowly cooled to 20°C, followed by rapid cooling to 5°C. The polymer was rigidized by treatment with a dilute solution of acetic acid. A triangular phase diagram was prepared representing polymer, water and total salt which included the Na_2HPO_4, the solvating agent, and Na_2SO_4, the coacervating agent. It was found that the amount of drug in the microcapsule had no appreciable effect on the particle size distribution of the microcapsule but did influence the release rates of the drug, suggesting that diffusion of the drug through the microcapsule wall is the controlling step. In order to obtain better utilization of the polymer, a technique was developed for the continuous addition of sodium sulfate during cooling of the system to produce microcapsules. The encapsulation process is able to produce microcapsules of varying drug-to-shell ratios by maintaining the polymer concentration and altering the amount of drug used. The rate of drug release increases as the drug content increases. In contrast, all batches of capsules plasticized by washing with a dilute solution of glycerin for a short period showed identical release rates, despite different drug contents; the authors now suggest that the release rate is controlled by dissolution of the drug in the microcapsule.

Another method of preparing microcapsules using the polymer cellulose acetate phthalate-containing pharmaceuticals with low water solubility was described by Milovanovic and Nairn (1986). Solutions were prepared by dissolving the polymer in dilute solutions of Na_2HPO_4 and heating to 60°C. Various quantities of the drug sulfadiazine, polyoxyethylene 20 sorbitan monooleate and, if necessary, a viscosity agent such as glycerin, Avicel pH 105 or hydroxypropyl methylcellulose were added to the polymer solution. The stirred suspension was added dropwise to the aqueous hardening solution of diluted acetic acid. A suitable viscosity of the solution to suspend the drug was obtained by using a 2.5% cellulose acetate phthalate solution. Addition of the above-mentioned viscosity agents to this solution did not alter the core:coat ratio, the particle size appreciably, or the percentage of drug incorporated, 81–94%, but the disintegration time was decreased when glycerin was used. The size of the microcapsules tended to increase, the core to coat ratio increased, and the disintegration time decreased as the amount of drug incorporated into the microcapsule increased.

Microcapsules of water-insoluble liquids such as vitamin A palmitate have also been prepared using cellulose acetate phthalate (Anon., 1988).

Gelatin

Phares and Sperandio (1964) showed that a number of insoluble particles, liquids and solids, could be encapsulated with gelatin using sodium sulfate as the coacervation-inducing agent. A phase diagram for the system gelatin, water and sodium sulfate was prepared to show the region of encapsulation.

As a result of the preparation of phase diagrams (Nixon *et al.*, 1966), suitable compositions within the coacervate region were selected for preparing microcapsules. Subsequently, an improved method for preparing microcapsules by simple coacervation methods using gelatin was accomplished by Nixon *et al.* (1968). The drug, sulfamerazine, was dispersed in either ethanol or 20% w/w sodium sulfate and added to the isoelectric gelatin solution. The mixture was stirred and maintained at 40°C. Both lime-pretreated and acid-processed gelatin were studied. After further treatment with the coacervating agent, the product was washed with isopropanol and hardened with a formalin-isopropanol mixture. This method produced the best results. In an alternative procedure for hardening the microcapsule, the product was cooled to 5 or 10°C, washed and dried; this method produced a cake. In a third method, the microcapsules were spray dried, but the product was not satisfactory because most of the drug was not encapsulated. The size of the drug particles to be coated did not hinder the coacervation process. It was found that encapsulation was successful if the drug particles were dispersed in the gelatin solution before coacervation or added to the system when coacervation was complete. The authors suggested that encapsulation can occur by two methods: the dispersed particles functioning as nuclei around which the coacervate drops form, or the coacervate droplets surround the drug particles. The recovery of the microcapsules was based on hardening the coacervate shell by dehydration. Isopropanol with its milder dehydrating effect compared with ethanol was more appropriate. The release rate of the drug was decreased with longer formalization time or thicker walls. Microcapsules prepared using ethanol provided a slower release than when they were prepared with sodium sulfate, which results in a more porous coat because of the salt's ability to hinder the hardening effect of isopropanol.

Nixon and Matthews (1976) made gelatin microcapsules by preparing a 5% solution of the polymer at 40°C and adding the coacervating agent, either 20% sodium sulfate or absolute ethanol. The core was added to some of the coacervating liquid and dispersed by ultrasonic vibration. The coacervate wall was then gelled by using a 30% ethanol in water or a 7% solution of sodium sulfate in water at a temperature below the gelling temperature of the coacervate; that is, below 12°C. Partial dehydration was accomplished by using two washes of isopropanol, a final wash with ethanol and finally heating the microcapsules to less than 60°C. The product was examined using a scanning electron microscope. Microcapsules produced by using either ethanol or sodium sulfate had no cracks or fissures. The surface of microcapsules produced using ethanol were smoother than those produced using Na_2SO_4. Surface folding of the ethanol-treated microcapsules was common and is associated with the formation of vacuoles within the alcohol coacervate droplets. The authors suggest that during recovery the vacuoles collapse and the wall material folds in on itself. Crystalline deposits on the surface of sodium sulfate-produced microcapsules were that of the salt. The authors suggest that microcapsules prepared by coacervation are formed by a process that involves the combination of several smaller microcapsules.

Water in oil emulsions have been encapsulated by gelatin using the coacervation process. For example, a concentrated solution of urea in water was prepared as a w/o emulsion with corn oil and hydrogenated castor oil. A solution of gelatin and the above emulsion were heated to 40°C and dispersed slowly in a stream into a solution of sodium sulfate at 40°C with stirring. After phase separation, the mixture was cooled, adjusted to pH 9.5 and treated with formaldehyde to harden the product (Heistand *et al.*, 1970).

The effect of ethanol, sodium sulfate and resorcinol on the induction period and some physical properties of gelatin coacervates has been studied by Zholbolsynova *et al.* (1971). Later Zholbosynova *et al.* (1988) investigated the influence of alcohol on the rheological properties of aqueous solutions of gelatin during the formation of coacervates. It was found that the viscosity of the coacervates increased with increasing concentration of the alcohol in the order methyl alcohol < ethyl alcohol < propyl alcohol. The strength of the coacervates prepared with ethanol as the coacervation agent increased with time, and was at maximum at an ethanol concentration of 13% v/v.

Nath (1973) investigated the influence of coacervation volume as altered by the temperature and the coacervating agent. It was found that as the temperature of coacervation increased in the system, gelatin, water, sodium sulfate, the volume of the coacervate increased. The volume decreased as the concentration of the coacervating agent, sodium sulfate, increased from 4.5 to 6.6%. The addition of hydrocolloid also altered the coacervation process. Dilute solutions 0.05–0.1% of carboxymethylcellulose reduced the growth of microcapsules during gelling. At higher concentrations, 0.1 to 1%, it increased the viscosity of the system. However, the capsules could not be filtered from the viscous liquid. The addition of polyvinyl pyrrolidone promotes flocculation and this interferes with coacervation.

Later, Nath and Shirwaiker (1977) studied the enhanced adsorption of atropine sulfate by kaolin in the presence of the coacervation-phase of gelatin, compared with either kaolin or the dried encapsulated form separately. The enhanced adsorption was attributed to the altered surface characteristics of the adsorbent in the gelatin–Na_2SO_4 system. Release of the drug from the coacervated kaolin product into simulated gastric or pancreatic fluid *in vitro* was considerably slower than that from the other two forms.

Simple gelatin coacervate systems have been used to enhance drug uptake by adsorption. Nath and Borkar (1979) prepared gelatin coacervates using ethanol as the coacervating agent. In three separate experiments, the amount of amphetamine bound by kaolin, gelatin coacervate and the kaolin gelatin coacervate system was studied. The authors suggest that the enhanced uptake of the drug by the coacervated kaolin results from successive layers of the coacervate phase providing new surfaces for drug deposition. Drug release in gastric and pancreatic fluid from the kaolin gelatin product follows first-order kinetics. Addition of surfactant to the dissolution fluid, Tween 20, Tween 80 or sodium lauryl sulfate, enhances drug release, which suggests that the drug material is bound by both the core material and the coacervate coat.

Coacervation of gelatin in the presence of surface-active agents has been investigated for a number of reasons by Ohdaira and Ikeya (1973) and Ikeya *et al.* (1974a), who encapsulated lypophilic materials or water-insoluble substances using gelatin and a quarternary ammonium salt, e.g. octadecyl-trimethyl ammonium bromide. The microcapsules were hardened with for-maldehyde in an alkaline solution to give independent microcapsules. Subsequently, Ikeya *et al.* (1974b) used an anionic surface-active agent, e.g. sodium lauryl sulfate, in the coacervation process to aid in the encapsula-tion of hydrophobic materials.

Two coacervate systems of gelatin–benzalkonium chloride and acacia-gelatin were prepared and analysed for the sorption of halothane. Signifi-cant halothane gas uptake was observed in the highly structured coacervate system (Stanaszek *et al.*, 1974).

Coacervation of gelatin has been promoted by the addition of poly(vinyl alcohol). The agglomeration of gelatin was attributed, by the authors Falyazi *et al.* (1975), to be the interaction of poly(vinyl alcohol) and water, which alters the solubility of gelatin and promotes coacervation.

In order to improve its surface properties, pyrvinium pamoate was encap-sulated with gelatin. Optimum results were obtained using a 10% gelatin solution at 50°C at a core to coat ratio of 2:1. Trivalent and divalent ions were effective in promoting coacervation when phase separation did not occur with NaCl. The addition of Tween 80 to the system before coacerva-tion produced microcapsules that contained larger amounts of drug than when the surfactant was added after phase separation (Kassem *et al.*, 1975a).

An inorganic polymer has been used to induce coacervation. Hoerger (1975) induced phase separation of gelatin using Calgon (sodium hexameta-phosphate) at 80°C using a lipophilic material as a core.

The stability of microencapsulated vitamin A and vitamin D concentrates in olive oil was not nearly as good as the non-encapsulated product stored under the same conditions, both in the presence of light and protected from light. It was suggested that the decreased stability was due to the porosity of the gelatin membrane which permitted light and moisture to reach the vitamins (Spiegl & Jasek, 1977).

Highly volatile liquids have been encapsulated with gelatin using sodium sulfate to effect coacervation. The microcapsules were then treated with isopropanol and formaldehyde. The product was further treated with stearic acid to prevent loss of the liquid cyclohexane. The microcapsules with a size range of 20–50 μm contain 75% of the volatile liquid (Spittler *et al.*, 1977).

Coacervates of gelatin and benzalkonium chloride have been prepared from 10% and 5% solution, respectively, by Takruri *et al.* (1977). These coacervates were compared with organic solvents with regard to the partitioning of four barbiturate salts and also the absorption of the barbiturates in the rat colon. The authors suggest that coacervation systems form a more realistic model for studying the absorption characteristics of drugs than do conventional organic solvent–water systems.

Madan (1980) studied the release behaviour of microencapsulated clofibrate, a liquid hypercholesterolaemic agent and related the data to the formation of the microcapsules. The drug fell from a capillary tube into a stirred, warm solution of gelatin type B. Then a 20% solution of sodium sulfate was added to promote the coacervation of the oil droplets. The product was poured into a 7% solution of sodium sulfate to gel the wall. Chilled isopropanol was added to dehydrate and flocculate the coacervate drops. The microcapsules were then hardened by immersion in a 10% solution of formaldehyde for up to 8 h. The process produced discrete, free-flowing particles of a uniform size (190 ± 10 μm). The dissolution of the microencapsulated drug in a 30% isopropanol solution at 37°C was studied. Several mathematical models were tested (square root, Langenbucher, cube root) but none yielded linear graphs. A close examination of the graphs showed four linear segments. The authors suggest that the matrix of the microcapsule differed from that proposed in the release of drug from solid matrices or from uniform non-disintegrating granules which tend to be homogeneous. The matrix appeared to be composed of various layers which exhibit different release characteristics.

The influence of glucose syrups and maltodextrin was studied by Marrs (1982). It was found that the inhibitory effect on gelation increases with the amount of high molecular weight oligosaccharides in the system. In addition, the properties of gelatin are modified by the composition of the starch hydrolysate.

Shchedrina *et al.* (1983) encapsulated dibunol by means of coacervation with gelatin solution. The stability of the microcapsules was investigated by determining such properties as bulk weight, friability and wearing properties after storage for 2.5 years at 20°C and 5°C. *In vivo* studies show the absorption of the oily drug was more uniform, continuous and prolonged compared with the oily liquid itself.

Nikolayev and Rao (1984) studied the effects of plasticizers on some physical properties of gelatin microcapsules prepared by coacervation. The microcapsules were prepared by treating a solution of gelatin at 50°C with a 20% solution of sodium sulfate. Oil coloured with Sudan III was added with stirring and the mixture cooled to 5°C to form the microcapsules. Microcapsules were also prepared by adding suitable amounts of plasticizer to the warm solution of gelatin prior to the addition of sodium sulfate. The resulting microcapsules were mono dispersed with a size range of 300–400μm. The surface was smooth, the wall material was uniformly distributed and the coat on the plasticized microcapsule was thinner than on non-plasticized product. As the concentration of the plasticizer, glycerol, sorbitol, propylene glycol or polyethylene glycol 400 increased, the percentage of gelatin deposition decreased and there was also a tendency for a decrease in wall thickness. Finally, as the concentration of plasticizer glycerol or sorbitol was increased, the time for 50% release of the oil, as determined by dye concentration, decreased in a linear manner.

A matrix formulation of small particles, encapsulated with gelatin, has been prepared by decreasing the pH, causing the drug to precipitate from solution, and simultaneously effecting coacervation (Frank *et al.*, 1985). Sodium sulfadiazine was encapsulated by this method by titrating a solution of the drug in water containing ethanol, sodium sulfate and gelatin with HCl. A white suspension of microencapsulated particles was formed, which was poured into cold Na_2SO_4 solution and then stirred at the bath temperature to effect gelling of the liquid gelatin microcapsule shell.

A matrix encapsulation formulation of small particles of a water-insoluble drug, felodipine, was prepared by dissolving the drug in a little polyethylene glycol 400 and adding to the solution a 2.5% gelatin solution containing Na_2SO_4 which caused precipitation of the drug and the formation of a coacervate around the fine drug particles. A solution of Na_2SO_4 was added to complete the encapsulation; all steps were carried out at 55°C. The wall was gelled by pouring the suspension into a cold Na_2SO_4 solution and hardened with formaldehyde (Brodin *et al.*, 1986).

Gelatin coacervates have been prepared in the annulus between rotating concentric cylinders. Coacervation was induced in the water by the addition of Na_2SO_4. The coacervate droplets showed a logarithmic, normal distribution and their size depended upon the rotation rate, residence time and pH (Yagi, 1986, 1987).

Coacervation and encapsulation of fat materials such as cosmetics using gelatin was achieved at 50°C, using a small quantity of sorbitol, effecting coacervation with carrageenan, followed by treatment with glutaraldehyde to form microcapsules. This product may then be treated with other polymers such as a mixture of polydimethylsiloxane and poly(vinyl pyrrolidone) for printing on paper (Fellows *et al.*, 1987).

Rozenblat *et al.* (1989) investigated the effect of electrolytes, stirring and surfactants in the coacervation and microencapsulation process using gelatin. Lime-pretreated bovine skin gelatin with a gel strength of 60 and 225 bloom and acid-processed porcine skin gelatin with gel strength of 175 and 300 bloom were used. The first step in the procedure was the addition of the core oleic acid and surfactant to an aqueous gelatin solution (8% w/v, pH 6.0–6.5) at 37°C, with stirring to effect emulsification. The second stage in the procedure was the encapsulation by adding a solution of 20% sodium sulfate. Finally a cool solution of sodium sulfate (7%) was added. The coacervation process was monitored by turbidity measurements. The microcapsules were observed and measured by using the microscope and a Coulter counter, or a computerized inspection system. It was found that coacervation is indifferent to the nature of the charge on the gelatin. However, an increase in bloom strength of the gelatin required smaller amounts of Na_2SO_4. These findings support the theory of Nixon et al. (1968). Experiments using different electrolytes could be divided into three groups. A number of fluoride salts were used to induce phase separation, for example MgF_2 and NaF. It was found that the effects of the electrolytes to induce phase separation increased with the charge density and the solubility of the electrolyte. The group was called phase separation inducers. Salts of polyvalent anions also belong in this class. A number of salts, e.g. $NaNO_3$ and NaI, require a greater amount of Na_2SO_4 to induce phase separation; these monovalent salts are known as chaotropic salts and have the ability to destabilize membranes. They decrease the energy required for solubilization and therefore increase the solubility of the gelatin. The efficiency of the chaotropic salts as inhibitors of coacervation decreases with an increase in the charge density. The inert salts do not induce phase separation and do not change the solubility of the polymer in water. Their charge density is between the previous two groups. The authors were able to encapsulate oleic acid in the presence of positively charged gelatin, non-ionic surfactants and anionic surfactants, but not in the presence of a positively charged surfactant. The inability to coat the oil drops in the presence of a positively charged surfactant was attributed to electrostatic repulsion. Encapsulation of the oil was not successful with negatively charged gelatin, but it could be encapsulated in the presence of non-ionic surfactants except Tween 20. These results disagree with those of Siddiqui and Taylor (1983). The control of stirring speed is most important during the cooling stage of the process, as it determines the microcapsule size. Prolonged stirring at stage two tends to cause an increase in aggregation.

Microcapsules of cholecalciferol were prepared by both simple and complex coacervation using gelatin A and gelatin B. Research of Sawicka (1990) shows that the properties of the microcapsules depend upon the coat to core ratio regardless of the type of gelatin or the coacervating process used. The size of the microcapsules, their dissolution in digestive juice, the coat to core ratio, the core content, and the rate of drug release were determined. With simple and complex coacervation methods, the optimum coat to core ratios were 0.25:1 and 0.5:1, respectively.

Coacervation with gelatin has been used to encapsulate drugs with an unpleasant taste. Ozer and Hincal (1990) encapsulated beclamide by simple coacervation by adding the drug to a stirred solution of gelatin at 40°C. Sodium sulfate solution was added slowly over a 35 min period. After cooling, decantation and washing with water, the microcapsules were hardened by the addition of a 75% w/v potassium aluminium sulfate solution at pH 4 and 7°C. Several other hardening agents were employed: formaldehyde, glutaraldehyde and isopropanol–aldehyde solutions. In order to improve flow properties, some of the microcapsules were dispersed in isopropanol 50% at 4°C containing Aerosil. The addition of alcohol during the preparation extracted some of the active ingredient, resulting in decreased beclamide content. Glutaraldehyde was found to be the best hardening agent. Aerosil tended to prevent the microcapsules from sticking together, in contrast to isopropanol. The mean size of the drug particle was 127.5 μm and after microencapsulation the mean size was 550 μm. The authors investigated some of the properties of the microcapsules, namely flow, consolidation, and the apparent and tapped densities. The release rates of the drug were found to be dependent on the type of gelatin and the method of hardening. Microcapsules prepared with no hardening agent had the fastest rate of release, those hardened with glutaraldehyde had intermediate rates whereas those hardened with an aldehyde and isopropanol mixture had the slowest release rate. The authors also prepared three types of tablet formulations: conventional, chewable and effervescent. Physical properties and dissolution of the drug from the tablets were also studied.

Nikolaev (1990) found that the physicochemical properties of gelatin microcapsules prepared by coacervation depended upon the polymer:core ratio and the treatment with formaldehyde. The particle size and the specific surface area of the microcapsules were influenced by the formaldehyde treatment. Properties of the final product, which contained norsulfazole as the model drug, were also dependent on the polymer density, the bulk mass and thickness of the microcapsule coatings when the number of drug particles increased.

Gelatin has been used to encapsulate natural and partially synthetic oils. The method developed by Keipert and Melegari (1992) enabled the preparation of microparticles that were approximately spherical and had a particle size of about 100–600 μm. It was shown that pH, the type of gelatin and additives to the gelatin system influence the characteristics of the microparticles, such as the mean diameter and surface, through the effects of viscosity and interfacial tension. The quality of the microparticles is also influenced by the characteristics of the core liquid, particularly the amount of unsaturated fatty acid.

Gelatin derivatives

A derivative prepared by the treatment of gelatin with succinic acid in alkaline pH was precipitated with Na_2SO_4 by lzgu and Doganay (1976). The heavier molecular weight fraction was separated and used to prepare microcapsules of sulfisoxazole. The microcapsules were hardened with alum and then with glutaraldehyde. Dehydration of the capsule wall was accomplished by using isopropanol and Aerosil. The drug release rate depended upon the degree of hardening and was found to be greater than from natural gelatin microcapsules.

Hydroxypropyl cellulose

Hydroxypropylmethyl cellulose and hydroxypropyl cellulose have been used as wall material for the preparation of encapsulated pharmaceuticals such as tocopherol acetate. The microcapsules were prepared by dissolving the wall material in water, along with dextran and effecting dehydration phase separation to give a product size of 100–300 μm diameter in the dispersion. The dispersion was then sprayed into fluidized silicone dioxide and dried to give microcapsules covered with silicone dioxide (Oowaki et al., 1988).

Methylcellulose and derivatives

Methylcellulose or derivatives have been employed as a wall material to encapsulate lemon oil (Takahashi et al., 1989). The oil was added to a 5% solution of methylcellulose to give an o/w emulsion, then a concentrated solution of hydrolysed starch as the coacervating agent was added to form the microcapsules, followed by dehydration with a 25% solution of NaCl and rinsing.

Polyethylene glycol

In a series of papers, Szretter and Zakrzewski (1984a,b, 1987a,b) encapsulated the vitamins riboflavin, thiamine nitrate, ascorbic acid and nicotinamide with polyethylene glycol (PEG) 6000 or 10 000 by heating, for example, a mixture of the vitamin and the polymer with paraffin oil, ligroin and Span 60. After cooling to room temperature, the product was washed with ligroin. The products showed improved stability with regard to humidity, air and light, depending upon the vitamin used.

Poly(vinyl alcohol)

Coacervates of poly(vinyl alcohol) have been prepared using colloidal silica as the core material (Iler, 1973, 1974). The coacervate was prepared from a dilute aqueous solution of the polymer and colloidal silica at pH 10. The pH was lowered to 2.6 with HCl and deaerated with a mixture of dodecyl alcohol and propyl alcohol, boiled in a vacuum and cooled to 25°C to form the coacervate. The maximum yield of coacervate was obtained when the ratio of colloidal silica to polymer in the coacervate was proportional to the particle diameter. For every nm^2 of colloidal silica surface, there were 2.5 hydroxy-ethyl chain segments. The coacervate consisted of silica particles whose surface was covered with a monomolecular layer of the polymer. The hydroxyl groups of the polymer were bonded to the SiOH group on the silica surface so that the hydrocarbon chain formed a hydrophobic coat.

Poly(vinyl methyl ether maleic anhydride)

Mortada (1981) prepared microcapsules of phenacetin using the polymer n-butyl half ester of poly(vinyl methyl ether maleic anhydride). The optimum conditions for coacervation were determined from a triangular phase diagram. Sodium acetate was used to dissolve the polymers in water and sodium sulfate was used to effect coacervation. The optimum conditions were polymer 0.5–3%, total salt 6.5%. During the microencapsulation of phenacetin, the coacervate drops tended to deposit on the surface of the drug suspended in the system. Gradual addition of sodium sulfate resulted in the continuous deposition of the coacervate onto the coated drug particle. As the core to coat ratio increased, the mean diameter of the microcapsules decreased slightly. However, the time required for the maximum amount of drug to be released decreased. Drug release was enhanced as the pH of the dissolution medium increased.

Poly(vinyl pyrrolidone)

Porous adsorbents such as activated carbon have been encapsulated with poly(vinyl pyrrolidone) to give improved selectivity. A dispersion of the adsorbent in a concentrated solution of the polymer was treated with a 30% solution of gum arabic, an incompatible polymer, and then coagulated with sodium sulfate and tannic acid to give a powder about 0.3 mm in diameter (Fukui, 1978).

Povidone and sulfathiazole microcapsules were prepared by suspending the drug in 2% solution of the polymer; a 20% solution of resorcinol was then added dropwise until complete separation was obtained. The product was then filtered and air dried to give a fine free-flowing powder. The authors Badawi and El Sayed (1980) also prepared a number of interaction products of the drug and polymer by different means and compared the various products by solubility and dissolution tests. The coacervate complex had higher solubility and a slightly higher dissolution rate than sulfathiazole. It resisted the action of dilute acids and alkaline which suggested a bonding between the polymer and the drug. A model is discussed to describe the dissolution profiles of the different systems prepared.

Starch

Starch microgels have been prepared by coacervation by Ohno and Higano (1992). Separate concentrated solutions of starch and poly(ethylene oxide) were prepared at 80°C, and then were mixed to form simple coacervates of starch by means of phase separation. After immediately cooling in an ice-cold water bath, spherical starch microgels were formed with an average size of 35 μm. The diameter could be increased, for example to 180 μm, by lengthening the incubation time, for example to 10 min.

Various polymers

Meshali *et al.* (1989a) encapsulated both oxyphenbutazone, acidic, and glafenine, basic, non-steroidal anti-inflammatory drugs with various water-soluble polymers. Carboxymethylcellulose, an acid polymer, was dissolved in water and the drug was added with stirring. This dispersion was then poured in a thin stream into a concentrated solution of aluminium sulfate for the oxyphenbutazone or calcium sulfate solution for the glafenine drug with stirring. After coacervation was complete, as noted by the presence of a clear supernatant, the product was filtered and dried. In a separate preparation, hydroxypropyl methylcellulose, a neutral polymer, was dissolved in toluene by heating at 40°C, the drug was dispersed with stirring and then the mixture was cooled. After filtration, the product was dried. In another preparation, chitosan (amino cellulose), a basic polymer, was dissolved in a dilute acetic acid solution and added dropwise to glafenine which was dispersed in a sodium hydroxide solution. After stirring for 1 h the polymer precipitated on the drug particles which were then washed with water and dried. Microcapsules of the two drugs were also prepared by fluidization technique. The percentage yield in the coacervation technique ranged between 68 and 94% and the actual amount of the drug present in the microcapsules ranged from 95.8 to 98% of the theoretical amounts. The microcapsules consisted of irregular aggregates thought to be due to polymer bridging caused by traces of polymer remaining in solution. The average diameter increased as the ratio of coating material in the microcapsule was increased. The dissolution rates were discussed in terms of the property of the drug, acidic or basic, and solubility of the drug as affected by pH and polymer concentration and the pH of the dissolution medium and the properties of the polymer. In a subsequent paper, Meshali *et al.* (1989b) found that all cellulose derivatives decreased the gastric ulcerogenic activity of the drugs.

COACERVATION-PHASE SEPARATION USING TWO WALL-FORMING POLYMERS SOLUBLE IN WATER

Acrylates

Acid and basic methacrylate polymers based on methacrylic acid and dimethylaminoethyl methacrylate were prepared and characterized by nuclear magnetic resonance spectrosocopy and dilute solution viscometry. Relations between pK_a, pK_b, pH, solubility and extent of ionization were determined by acid–base titration. Complex coacervates of these high charge density polymers were prepared and both yield and water content determined. Short-term cell viability using erythrocytes in these microcapsules was shown. The coacervates may be used for prostheses for organ transplant (Wen *et al.*, 1991a).

In a second paper, Wen *et al.* (1991b) reported on the ionizable group content, structure, molecular weight, solubility, solution behaviour and efficacy of ionic complex formation through complex coacervation for a range of sparingly soluble synthetic weak polyelectrolyte polymers with low charge content based on hydroxyalkyl methacrylate. Selected polymers containing methacrylic acid and dimethylaminoethyl methacrylate showed promise as forming pairs for the entrapment of mammalian cells. The solubility of basic polymers was enhanced through quaternization of the *N*-methyl group. The survival of guinea pig erythrocytes was indicated.

Albumin–acacia

Coacervates of serum albumin and gum arabic are unstable with respect to size. Studies of these static coacervates, that is coacervates in which no chemical reactions are occurring, showed that the turbidity increased with time to a maximum, then after a period of time, the turbidity decreased.

The turbidity was dependent upon the initial concentration of the coacervate forming system. Furthermore, based on light-scattering measurements, the coacervates were not stable during the period in which the turbidity measurements remained constant, because the coacervates coalesced into larger droplets (Gladilin, 1973).

The coacervation of albumin–acacia has been studied by Burgess *et al.* (1991a). Microelectrophoresis studies were used to determine the optimum pH condition for complex coacervation to occur. At pH 3.9 the polymers carry equal and opposite charges. The coacervate yield was maximum at an ionic strength of approximately 10 mM. The coacervate yield decreased at both lower and higher ionic strengths. High ionic strength affects the charge carried by the polymers through a screening effect of the counter ions; thus, the attraction of one polymer for the other is decreased. The decrease in coacervate yield with decreasing ionic strength is not predicted by a number of theories. This phenomenon can be explained when the configurations of the molecules are taken into account. Highly charged molecules may exist in a rod-like configuration, rather than a random coil, and thus act as specific sites in a distributive manner. The high viscosity of the coacervate phase at optimum pH values for maximum coacervation makes it difficult to disperse the coacervate which is required for the formation of small spherical microcapsules.

In a second paper Burgess *et al.* (1991b) further investigated the complex coacervation of bovine serum albumin and acacia. Maximum coacervation was predicted to occur at pH 3.9 where both polyions carry equal and opposite charges. The optimum ionic strength for maximum coacervation yield was 10 mM. At pH 3.9 the viscous coacervate phase could not be emulsified into the equilibrium phase; however, at pH 3.8 and 4.2, microcapsules could be successfully prepared. Particle size of the microcapsules decreases slightly as the stirring speed increases.

Albumin–alginic acid

The albumin–alginic acid complex coacervate process has been investigated by Singh and Burgess (1989). Maximum coacervation was obtained at pH 3.9, which is the electrical equivalence point for this pair of polymers where alginic acid and albumin carry equal and opposite charges. The optimum ionic strength for maximum coacervation is low in this system which is in agreement with the microelectrophoresis data. At low ionic strength (5–50 mM), there is a strong interaction for coacervation to occur. At high ionic strength coacervation is suppressed, while at very low ionic strength a precipitate is formed owing to the exclusion of water. Complex coacervation was observed at 0.05 and 0.5% w/v total polymer concentration.

Above 0.5% w/v only precipitation occurred. Opalescence noted in the system suggests that spontaneous aggregation takes place, and enhanced coacervation upon temperature reduction indicates a slow rearrangement of aggregates to form the coacervate phase. The albumin-alginate system fits the Veis–Aranyi (1960) theory which states that coacervation occurs in two steps spontaneously: aggregation by ion pairing followed by a rearrangement of the aggregates to form a coacervation phase. Coacervation is limited, compared with other polypeptide-polysaccharide systems such as gelatin-acacia, due to the occurrence of precipitation rather than coacervation if some of the conditions are not optimum. The albumin-alginic acid coacervation is very viscous and this makes the system unsuitable for preparation of microcapsules.

Albumin–dextran

The interaction of diethylaminoethyldextran with bovine serum albumin has been investigated by Gekko et al. (1978). It was investigated at the pH range above the isoelectric point by turbidity and electrophoresis. The coacervate was affected by pH, ionic strength and the weight ratio of the compounds. The interaction between the two compounds was attributed to electrostatic charges and was further strengthened by hydrophobic effects. The localized charge distribution model of protein polyion interaction is supported.

Alginate–chitosan

Capsules consisting of precipitated chitosan, a chitosan–alginate interphasic membrane and calcium alginate layers were prepared. The wall properties such as mechanical stability and thickness could be altered by buffer treatment and by partial substitution of alginate with propylene glycol alginate (Knorr and Daly, 1988).

Dextran–polyethylene glycol

A microemulsified polyethylene glycol–dextran coacervate used to dissolve haemoglobin. Ecanow et al. (1990) indicated that this system mimics whole blood with regard to oxygen-carrying properties and can sustain life when transfused into an animal.

Gelatin–acacia

Luzzi and Gerraughty (1964) used gelatin and acacia to encapsulate various oily liquids. After effecting coacervation, first at pH 6.5 at 50°C, then at pH 4.5, formaldehyde was added and the mixture cooled to 10°C and then the pH was adjusted to 9.0 to rigidize the coacervate droplets. The permeability, as shown by extraction studies of the gelatin acacia shell, is not affected by oils with different saponification values ranging from 2.5 to 12. However, the amount of oil extracted increases as the acid values of the oils increase. The incorporation of the oil-soluble surfactant sorbitan trioleate 85 (Arlacel 85) interferes with the encapsulation process and gave erratic results upon extraction. Water-soluble surfactants, polysorbate 80 and sodium oleate appeared to prevent proper encapsulation. It was suggested the surfactant may compete with the gelatin–acacia complex at the oil–water interface.

In a later study Luzzi and Gerraughty (1967) encapsulated a number of different solids with gelatin–acacia, for example pentobarbituric acid, and then determined the quantity extracted in gastrointestinal fluids. They found that the quantity of solid extracted did not change very much over a 2.5 h period when the pH for inducing coacervation was varied. The minimum amount of drug extracted occurred when the starting temperature for the coacervation was 37°C. It was suggested that this was related to the fluidity of the coacervate and the solubility of the drug. The retaining power of the capsules increased as the ratio of core to coat decreased. It was suggested that this may be due to multiple droplets, thickness of the wall or the attraction of empty capsules to filled capsules.

The effects of a number of variables on the gelatin–acacia system were studied by Dhruv and colleagues (1975). An acid type gelatin with an isoelectric point of approximately 8.5 to 8.7 and acacia were separately dispersed and dissolved in water. The pH of each solution was adjusted to 6.5 with a concentrated solution of sodium hydroxide. The solutions were mixed and adjusted to specific pH values. The mixture was allowed to stand for 3 h and the coacervate volume was then measured. This system yielded liquids which have a low viscosity and thus it is easy to measure the coacervation volume. At a given temperature the volume of the coacervate increased as the pH increased from 3.5 to 4.5 and the maximum volume was obtained at an equilibration temperature of 40°C. Microscopic observations showed that droplets tended to be smaller at higher temperatures. It was found that there was a linear relationship between the acacia:gelatin ratio and pH values for maximum coacervation volume and the final pH. Finally, it was found that at a given pH value the coacervation volume increases to a maximum at 4% total concentration and then decreases abruptly; this was attributed to an increase in the number of inorganic ions present and possibly an increase in the viscosity of the system.

In a study to compare the release rates of sulfathiazole by microencapsulation by complex coacervation, Kassem and El Sayed (1973) used gelatin in combination with acacia, sodium alginate or pectin. It was found that the release follows a first-order process and the rate of release was in the order gelatin–alginate > gelatin–acacia > gelatin–pectin. Furthermore, the amount of drug release increased with a decrease in particle size in all systems.

In a subsequent paper Kassem and El Sayed (1974) investigated a number of factors that affected the release of sulfathiazole. They found that the starting pH for coacervation (6, 6.5, or 7) did not affect drug release after 2 h, but the final pH (8.5, 9.0 and 9.5) enhanced release into gastrointestinal fluids. The release was increased when the starting temperature was raised from 36 to 44°C and as the core to coat ratio increased, and when 1 or 10 ml of formaldehyde were used rather than 5 ml.

The reaction of glutaraldehyde with gelatin–gum arabic coacervate gels has been studied by Thies (1973). Both acid and alkali precursor gelatins were used to form coacervates. All gels had a consumption of 0.3 to 1.6 mmol of aldehyde per gram of gelatin. The consumption of glutaraldehyde by acid precursor gelatin acacia coacervate increases with gelation temperature 4–28°C due to changes in the gel structure. The reaction of the gels with the aldehyde causes insolubilization as a result of intermolecular cross-linking. Most treated gels were at least 85% insoluble at 55°C in phosphate buffer after 4–28 days of extraction. Gum arabic has little tendency to react with the glutaraldehyde, but is entrapped in the cross-linked structure.

Coacervate systems such as gelatin–acacia and gelatin–benzalkonium have been investigated in terms of their uptake of halothane. It was found that the coacervate took up considerably more halothane gas than did the dissolved or broken coacervate system. Stanaszek et al. (1974) suggested that the increased uptake indicates the presence of a highly structured, nonpolar system similar to surfactant micelles in a polar medium.

Optimal conditions for the preparation of gelatin–gum arabic films from coacervates have been investigated by Palmieri (1977). Various concentrations of gelatin and acacia were mixed at various temperatures and pH values. The pH was decreased to the 3–4.5 range and formaldehyde added. The filtered resuspended product was treated with isopropanol solutions and the product dried on a teflon coated sheet giving, on average, a film thickness of 95.61 μm.

Takenaka et al. (1979) related the wall thickness and amount of hardening agent to the release characteristics of sulfamethoxazole. The Higuchi model was used to interpret the release characteristics of the drug and linear correlations were obtained up to 60–80% release when water was the dissolution medium. The release rate decreased with increasing wall thickness of the microcapsule. An increase of formalization also delayed the release rate. Diffusion coefficients through the microcapsule wall were found to range between 1.63 and 283 $\times$ 10^{-9} cm^{2}s^{-1} and decreased with the coacervate pH and with increased amounts of hardening agent used. Tortuosity increased with an increase in coacervate pH and an increase in the amount of hardening agent used.

Sulfamethoxazole microcapsules were prepared by Takenaka *et al.* (1980a) using gelatin and acacia solutions, and by adding the drug to the acacia solution. The solutions were mixed at 50°C and treated with acetic acid until the desired pH (2.5–4) was achieved. After cooling to 5°C and washing, the coacervates were hardened with formaldehyde. As the pH increased, the particle size decreased. The particle size of formalized microcapsules was larger than that of the unformalized ones because formaldehyde prevents shrinking of the microcapsules during the dehydration and drying process. Unformalized microcapsules had a smooth surface in contrast to the wrinkled appearance of the formalized product. Spray dried microcapsules had folding and invagination. The optimum pH for coacervation was 3.5 and this produced the highest core content (77.5%). Spray drying of the product changed the drug into the amorphous form, as studied by X-ray diffraction.

Takenaka *et al.* (1980b) found that the zeta potential increased and then decreased with increasing amounts of formaldehyde and this was attributed to denaturing of the gelatin and release of acacia from the microcapsule wall. The zeta potential of spray dried microcapsules showed that the wall was denatured during the process.

Further studies by Takenaka *et al.* (1981) indicated that coacervation occurred on the surface of sulfamethoxazole particles with adsorbed acacia. They also showed that the coacervate droplets were deformed to an ellipsoid in an electrical field. This phenomenon, the Buchner effect, showed that the coacervate wall was flexible. The denaturation of gelatin by formaldehyde did not occur at the coacervation stage, but during succeeding drying processes which completed the hardening of the microcapsules. Spray drying also denatured the gelatin.

Takeda *et al.* (1981) encapsulated indomethacin in soybean oil by preparing a coacervate of gelatin and acacia and hardening the microcapsules with formaldehyde at 5°C for 24 h. Sodium hydroxide was not used in the hardening process as indomethacin is degraded in its presence. After repeated washing with water, no formaldehyde could be detected. The yield of product in terms of the original amount of drug was greater than 80%.

The microcapsules dissolved more slowly than indomethacin. Both the microcapsules and the soybean oil suspension gave higher and more prolonged serum concentrations than the drug alone.

Indomethacin in the form of a paste in a water-miscible liquid, polyhydroxyalkane, was encapsulated with a coacervate of gelatin-acacia. A 40% formaldehyde solution was used to harden the walls and, after treating with isopropyl alcohol, gave a free-flowing product. The microencapsulated indomethacin reduced the gastrointestinal irritation compared with uncoated drug (Rowe and Carless, 1981).

A somewhat different method of preparing microcapsules of an oil, namely liquid paraffin containing a dye, was patented in Japan (Shionogi and Co, 1982). An emulsion was prepared using liquid paraffin and an 11% gum arabic solution. Emulsification continued until the particle diameters were 30 to 50 μm. Then a solution of acid-treated 11% gelatin was added and a 5% solution of polyethylene oxide which effected phase separation. After dilution with water the pH was adjusted to 7.7, cooled to below 10°C and treated with formaldehyde to form microcapsules.

An elaboration of the above method was described by Jizomoto (1984). Briefly, an emulsion of liquid paraffin is prepared using acacia or other anionic polymers such as carboxymethylcellulose or ethylene–maleic anhydride copolymer. The gelatin solution (11%) and the non-ionic polymer, for example, polyethylene glycol 6000, either in solution or in flake or powder form are added. The pH is adjusted with a solution of NaOH and the mixture is cooled. The mixture is subjected to a hardening treatment using diluted glutaraldehyde or formaldehyde. The product is then filtered and washed. The authors indicate that the addition of a small amount of polyethylene glycol or polyethylene oxide to the system of gelatin acacia allows microencapsulation to proceed over an expanded pH range of 2–9 and spherical microcapsules can be obtained. Other non-ionic polymers were also investigated such as dextran or poly(vinylpyrrolidone), but gave coacervate systems of higher viscosity in the cooling process which made it difficult to prepare good microcapsules. The authors attribute the spherical shape of the microcapsules to the appropriate viscosity of the mixture which was affected by both the molecular weight and the concentration of the non-ionic polymer. Jizomoto (1985) further described the value of polyethylene oxide or polyethylene glycol in an aqueous system of gelatin with or without acacia for a wide pH range of 2.5–9.5. The expanded pH range makes both processes more useful for drugs which may be water-soluble or have stability problems in certain pH regions. Further information on the effect of molecular weight on the induction of phase separation was discussed in the theory section.

Indomethacin was also encapsulated using gelatin and acacia with and without hydroxypropyl celluose by Ku and Kim (1984). The rate of drug release and the drug content of the microcapsules decreased as the amount of wall material increased. The drug content was lower in microcapsules which contained the added polymer. The release rate of the drug from microcapsules with a 1:2 drug to matrix ratio was delayed as the formaldehyde treatment time was increased.

The release of nitrofurantoin encapsulated with gelatin–acacia coacervates was studied by Mesiha and El-Sourady (1984). The plain powder gave the fastest rate of release, followed by the directly encapsulated drug and then an encapsulated paraffin oil suspension of the drug. All samples showed a fast initial release rate followed by a slower sustained release with the encapsulated systems. As the viscosity of the oil increased, the release rate decreased.

A gelatin–acacia coacervate has been used by Noro et al. (1985) to prepare microcapsules of activated charcoal. The coacervates were gradually hardened with formaldehyde solution. Then the polymers were crosslinked by raising the pH above 9 using sodium hydroxide. The mixture was stirred at 50°C for varying times from 15 to 240 min. Adsorption studies of creatinine and components of higher molecular weight were carried out using coated microcapsules as the absorbent. As a result of the formaldehyde treatment, a stable semi-permeable membrane on activated charcoal was formed and the adsorption rate of creatinine was controlled by changing the cross-linking time. The substances with high molecular weight such as nutrients and enzymes had difficulty penetrating the membrane. It was suggested that administration of the encapsulated activated charcoal would be useful as a supporting technique in the treatment of patients with renal failure.

Indomethacin has been encapsulated by suspending the powder in rape oil and effecting coacervation with gelatin–acacia. The bioavailability from the microcapsule was high and showed a prolongation of action (Lu et al., 1986).

The preparation of microcapsules by means of complex gelatin–acacia coacervation has been facilitated by the use of ionizing colloids, ionic surfactants or ionizing long-chain fatty acids, added prior to the dispersion of the core (Ninomiya, 1986).

In a study of the microencapsulation of oils such as paraffin, lemon and orange, Arneodo et al. (1986) determined various interfacial tensions. The interfacial tension between the oils and water > oil and the supernatant > oil and coacervate. The oils showed two types of interfacial behaviour with respect to the aqueous solutions. In the first type, a decrease of interfacial tension occurred up to the attainment of pseudo-equilibrium and no chemical modification occurred at the interface. In the second type, the interfacial tension decreased to zero as a function of the temperature and the oil, and in this case a physicochemical modification occurred at the interface. Gelatin and various anionic substances were used such as Calgon 206, gum arabic and sodium alginate in the experiments.

Arneodo and colleagues (1987, 1988a,b) investigated a number of physicochemical properties of complex coacervates prepared from gelatin and anionic polymers. The solid contents and viscosities of the systems gelatin–gum arabic, gelatin–sodium polyphosphate, and gelatin–sodium alginate, were assessed. The solid content of the gelatin–gum arabic did not change much with the temperature, but the solid content of the other two systems decreased as the equilibration temperature decreased. The high viscosity of the last system was attributed to the high ionic interaction of the oppositely charged colloids. The interfacial tension of coacervates of the above three systems and four citrus oils were determined by the Wilhelmy plate method. The initial values were less than 8 mJ m^{-2} and decreased with time. At 40–50°C the interfacial tension decreased within 6 h to a value too low to measure. Further studies indicated that some of the constituents of the oil dissolved in water and were at least partly responsible for the decrease in interfacial tension.

Various types of surface active agents, anionic, cationic and non-ionic, have been shown not to have an appreciable effect on particle size of microcapsules prepared by coacervation using the gelatin–acacia system (Duquemin and Nixon, 1986). However, the amount of the core, phenobarbital, incorporated depends upon the type and concentration of the surfactant. The best conditions for encapsulating the drug were 2% w/w colloid at a stirring speed of 180 or 250 r.p.m. and the inclusion of 0.025% w/v cetrimide. High concentrations of the surfactant decreased the encapsulation of phenobarbital owing to a decrease in interfacial tension, and also steric and electrostatic effects caused by surfactant adsorption onto the coacervate drops and core.

Nixon and Wong (1989) evaluated the permeation of three compounds through polymeric membranes as a model for the release of drug from gelatin–acacia microcapsules. The membranes were prepared by mixing equal volumes of 2% gelatin and acacia solutions at 41°C with stirring and adjusting the pH to 4.0–4.2, then adding a formaldehyde solution. Three methods of casting were employed:

1. Without prior removal of the equilibrium fluid.
2. Removal of the equilibrium fluid by centrifugation at 20°C.
3. Cooling to 4°C and removal of the equilibrium fluid by filtration and then resuspension of the coacervate with water.

Each of the coacervates were then poured into steel dishes and dried to form the film. Methods 1 and 3 produced films which had incomplete fusion of the coacervate droplets, as seen by scanning electron microscopy. In contrast, the film formed by slow phase separation, method 2, permitted good fusion of the droplets and resulted in a smooth surface. These films were similar to their corresponding microcapsules which are usually non-porous. The swelling of the films depended on the use of the cross-linking agent and indicated that swelling was complete after about 8 min for the cross-linked film, whereas with the non-crosslinked film, swelling continued to increase slowly for 60 h. The estimated permeability of the microcapsule wall was approximately 10^2–10^4 less than that of the film. This was attributed to difference in cross-linked density between the microcapsule wall and the cast film. It was found that gelatin–acacia coacervate does not slow the release of the drug core and, with N-7 theophylline acetic acid, a faster dissolution was obtained compared with the unencapsulated drug.

Burgess and Carless (1986a) studied the microelectrophoretic behaviour of gelatin–acacia complex coacervates and gelatin–gelatin complex coacervates and found that it was identical to that of equivalent mixtures of these polyions adsorbed onto a colloid carrier, namely silica with a geometric mean diameter of 2.7 μm. The charge carried by gelatin–acacia coacervates is not affected by encapsulated indomethacin particles which indicates that the drug is completely encapsulated and is likely not to be present in any significant amount in the capsule wall. This is in contrast to the properties of gelatin–gelatin microencapsulated indomethacin which suggests that the drug particles are associated with the capsule wall and thus produce a change in electrophoretic mobility.

Huttenrauch (1986) prepared gelatin–acacia microcapsules in the presence of structure breakers and structure formers. The latter, such as sorbitol, fructose and sucrose were found to promote larger microcapsules, helical structure formation and the agglomeration of gelatin particles. Structure breakers such as urea, methylacetamide and nicotinamide showed the opposite effect.

Pal and Pal (1986) investigated the complex coacervates of sulfamethoxazole with gelatin to develop a controlled release dosage form. Rigid microcapsules were obtained at pH 5.2 at 40°C with a drug polymer ratio of 1:1. It was found that a 28.7% v/v solution of glycerin and chilled propyl alcohol were necessary for the production of discrete spherical microcapsules.

Sulfamethoxazole microcapsules were prepared by coacervation with gelatin and acacia and a study of their micromeric properties was made by Pal and Pal (1987). The number of microcapsules per gram, the density of the wall material and that of the microcapsule increased with decreasing capsule size. The wall thickness decreased with decreasing capsule size and was inversely related to the square root of the number of microcapsules present. Infra-red and X-ray analysis showed that the drug did not complex with gelatin. Microcapsules treated with formaldehyde had thicker walls and gave a more controlled release than unformalized microcapsules. A linear correlation between the wall thickness and the *in vitro* T50% release was observed.

In a subsequent paper, Pal and Pal (1988) prepared gelatin–acacia microcapsules by a standard method, but a 28% w/v solution of glycerol was used as a plasticizer. The product was treated with formaldehyde, water and isopropanol. It was noted that the wall thickness of the microcapsules decreased and that the density of the wall increased with decreasing size. This was related to the volume fraction of the pores in the wall, which decreased with decreasing capsule size. The kinetics at the early stages of release were zero order. The apparent diffusion coefficient of the drug, sulfamethoxazole, decreased with decreasing capsule size. The release of the drug from the microcapsule was faster at pH 7.2 than at pH 1.2, although the diffusion coefficient was lower. This phenomenon was explained by taking into account the solubility of the drug, its pK_a, and the volume fraction of the pores in the membrane. The decrease of the apparent diffusion coefficient of the drug through the wall at sink conditions at pH 7.2 is due to the higher concentration gradient because of the drug's higher solubility compared with the value of pH 1.2.

Sage oil was encapsulated with a gelatin–acacia coacervate to give a particle size of 50 to 500 μm with a wall thickness of about 0.5 μm and the percentage of sage oil encapsulated was 94.7%. The antimicrobial activity of the sage oil and the encapsulated product were similar; however, lower activity of the encapsulated oil against fungi was seen (Jalsenjak *et al.*, 1987).

Ion-exchange resin complexes of ester prodrugs of propranolol and the drug itself have been encapsulated with gelatin–acacia coacervates to produce microcapsules that extended the time for 50% release from 25 to 100 min. The rate of release decreased as the ratio of core to coat decreased. Using the same coacervation procedure a double coat was formed and extended the time for 50% of the drug release to 4 h or more. Despite delayed release profiles, the authors, Irwin *et al.* (1988) found that release follows particle diffusion models.

Spiegl and Viernstein (1988) investigated the effect of various ratios of gelatin and acacia gum, types of gelatins and pH on the coacervation process. The best yield of the coacervate, 92%, with a minimum degree of hydration, was obtained at a pH of 3.9 and a gelatin–acacia ratio of 1.5:1 using a Stoess type of gelatin with a bloom strength of 250–260. This system was used to encapsulate griseofulvin.

Multicore microcapsules with a diameter of 10–50 μm have been prepared by using polymer, surfactants and coacervation with gelatin and gum arabic. An emulsion prepared from paraffin wax, 10% aqueous gelatin solution and emulsifier was added to a 10% solution of gum arabic at 60°C and mixed with Disparlon 1860 (polyaminoamide polyester salt) at 40°C, treated with formaldehyde at pH 4 and 20°C and then heated to 50°C, changing the pH to 9 to give the product (Yoshida *et al.*, 1989).

Improved vitamin A stability in a water-in-oil type emollient lotion has been prepared by Noda *et al.* (1989). The vitamin was added to a component of liquid paraffin and cetyl isooleate and this mixture was encapsulated with gelatin–acacia and hardened with glutaraldehyde and subsequently incorporated into an emollient lotion.

Microcapsules formed by the coacervation technique from gelatin and acacia, using glutaraldehyde, were prepared by Noda *et al.* (1992) for cosmetic purposes. The breaking strength of the microcapsule was controlled by mixing a liquid and solid oil component in a suitable ratio. Chemical stabilization of vitamin A palmitate, for example, was improved compared with the emulsified substance. Further improvement in stability was achieved by increasing the wall thickness or by using polylysine to modify the gelatin film.

In order to obtain a very prolonged release of a pheromone analogue, Omi *et al.* (1991) first prepared a liquid mixture of white beeswax and 2-ethylhexylacetate which was added to a dilute solution of gelatin stirred at 250 r.p.m. at 70°C. After 5 min the mixture was cooled rapidly and filtered. These wax particles were then microencapsulated by complex coacervation with gelatin–acacia and treated with formaldehyde at pH 9 for 2 h. As the release of the pheromone analogue was too fast for the intended purpose, an alternative method of encapsulation was performed. In this method, the wax particles were dispersed in a 20% aqueous gelatin solution and this suspension (s/w) was redispersed in 200 g of liquid paraffin to give a (s/w)o product. After 30 min agitation at 250 r.p.m. a w/o emulsion composed of an aqueous formaldehyde and liquid paraffin was added to the (s/w)o dispersion. Cross-linking was allowed to proceed for a few hours, then the microcapsules were filtered, washed with benzene, cold water and ethanol. The microcapsules prepared using the complex coacervate method contain only a single core and the diameters of different batches ranged from 169 to 338 μm and the thickness of the wall was about 1.5 μm. The products showed a rapid initial release and were exhausted after only 1 week. The product produced by the multiple emulsion produced microcapsules of average diameters ranging from 618 to 1366 μm and contained a number of wax particles ranging from 10 to 70. The release profile was altered considerably and although the initial release rate was still not satisfactory, 60% of the pheromone still remained after 10 days, at which time release rate became almost constant. The authors estimate that it would take 47 days for 50% of the chemical to be released. Mass transport parameters such as capacity coefficients and diffusion coefficients were determined using a two-stage model. It was assumed that the mass transfer resistances dominate the transport process for the chemical; one was located in the wax particles and the other through the gelatin wall.

Peters *et al.* (1992) investigated the effect of bloom grade and isoelectric points of gelatin on the complex coacervate with acacia and the microencapsulation of theobromine. It was found that the electrical equivalence pH of gelatin and acacia were in the same range as the maximum coacervate volumes. The pH range in which the most coacervate formed for high bloom grade gelatin was smaller for alkali-processed gelatin than for acid-processed gelatins. A similar small pH range was observed for low bloom grade gelatin of the acid type, compared with high bloom grade of the same type. The total amount of complex coacervate increased and the relative content of theobromine decreased for gelatin with high bloom grade. Microcapsules prepared with a high bloom grade gelatin were irregularly shaped and showed poor flow characteristics; however, no difference in theobromine release profiles from various microcapsules was noted.

Gelatin–alginate

Wajnerman *et al.* (1972) investigated the coacervation between gelatin and sodium alginate using turbidity measurements at pH 3.5–4.5. The formation of electrically neutral complexes of the two polymers was postulated as the first step in coacervation with subsequent association with sodium alginate.

Cholecalciferol solution in peanut oil was microencapsulated by three methods (Sawicka, 1985): (a) simple coacervation with gelatin type B and Na_2SO_4; (b) complex coacervation with gelatin and sodium alginate; (c) complex coacervation with gelatin and cellulose acetate phthalate. The content of the microcapsules was approximately $600\,000\,IU\,g^{-1}$ and the dissolution half-time was 300–400 min. The products produced by methods (b) and (c) were insoluble in gastric juice and soluble in intestinal juice, while the product produced by method (a) was soluble in both.

Benzodiazepines have been encapsulated in a gelatin–alginate coacervate, then treating the product with tannic acid and with silica to produce a cake. The cake was dried in a fluidized bed system and the silica was removed (David *et al.*, 1988).

Gelatin–arabinate

Santamaria *et al.* (1975) investigated the coacervation phenomenon between gelatin and potassium arabinate. It was found that the yield of the coacervate was greatest at pH 3.5–4.1 and when potassium arabinate was in excess compared with gelatin. The coacervation decreased in the presence of potassium halide and the effect depended mainly upon the salt concentration and also on the ionic radius of the anion. Theoretical and mathematical considerations were used to interpret the phenomenon.

Gelatin–bacterial polysaccharide

Chilvers *et al.* (1988a,b,c) have described the encapsulation of sunflower, paraffin oils and aluminum particles by means of complex coacervation using gelatin and an extracellular bacterial polysaccharide. It was found that coacervation occurred only at the pH range of 3.0–4.5. The product was treated with glutaraldehyde and washed with isopropanol.

Gelatin–Carbopol

El Gindy and El Egakey (1981a,b) investigated the coacervation of gelatin A or B and Carbopol 934, 940, or 941 (source: B. F. Goodrich Chemical Co.). The two polymers were dissolved separately at 40°C, and the gelatin solution was added to the stirred Carbopol solution and various parameters were altered to investigate the system. Best results were obtained with gelatin type A at a pH of 6.8 with Carbopol 941 and the optimum ratio of Carbopol to gelatin was 1 : 10. An increase in total colloid concentration up to 1.1% w/v resulted in a parallel increase of sediment weight. At higher concentrations the sediment weight was less pronounced. Stirring at 300–350 r.p.m. gave almost spherical uniform coacervates with an average diameter of 59 μm.

In a subsequent paper, El Egakey and El Gindy (1983) found that glycerol in 20–33% v/v added after coacervation, produced smooth, spherical coacervates and if glycerol was added to the Carbopol solution prior to coacervation a coarser product was formed. The addition of glycerol rendered the microcapsules less coherent and reduced their adhesion to glass. The effect of increasing the concentration of formaldehyde was to increase the sediment volume of the coacervate. Formaldehyde-treated microglobules were treated with various volumes and concentrations of alcohols. The flocculation and sedimentation efficiency showed that 2-propranol at 60% concentration gave the best results. Microcapsules of sulfadiazine encapsulated with the gelatin–Carbopol coacervate were also prepared. As the coat to core ratio increased, the percentage of drug encapsulated increased, and the average size increased from 78 to 136 μm.

Gelatin–carboxymethylcellulose

Koh and Tucker (1988a) characterized the sodium carboxymethylcellulose-gelatin complex by adding the sodium carboxymethylcellulose solution to the gelatin solution at 40°C with stirring and allowing it to stand for 10 min before evaluation. Maximum coacervation as determined from maximum deviation from additive viscosity occurred at pH 3.5 and 30% sodium carboxymethylcellulose. Around this pH, carboxymethylcellulose is negatively charged and gelatin is positively charged, resulting in strong electrostatic behaviour and hence, complex coacervation. At a pH range of 5.0–7.0 positive deviation from additive viscosity behaviour occurred, but coacervation did not occur. The change in viscosity and complex coacervation is explained in terms of ionization and the folding of the colloid. For example, at pH 5, the isoelectric point of gelatin and the anionic carboxymethylcellulose, a negatively charged soluble carboxymethylcellulose–gelatin complex forms and, as a result, electrostatic repulsion leads to unfolding of the complex and the viscosity shows a large positive deviation. The pH range for complex coacervation was found to be 2.5–4.5, as observed by turbidity measurements. The authors suggest that coacervate wet weights and volumes cannot be used to predict optimal coacervate conditions due to a change in coacervate morphology with mixing ratio.

   In a second paper, Koh and Tucker (1988b) determined the chemical composition of the coacervate and equilibrium fluid phases of the sodium carboxymethylcellulose–gelatin coacervation complex. The coacervate batches were prepared at 0.75 and 2% total colloid concentration at pH values of 3.0, 3.5 and 4.0 and a range of sodium carboxymethylcellulose compositions of 10–60%. The colloid mixing ratio at which the peak coacervate yield occurred varied with the pH. Low viscosity and high viscosity grades of sodium carboxymethylcellulose gave similar results. Phase diagrams of the three components, water, gelatin and sodium carboxymethylcellulose at different pH values were prepared. Changes in the colloid composition of the complex coacervate and equilibrium fluids of isohydric mixtures as a function of the sodium carboxymethylcellulose mixing ratio were determined. The authors concluded that the sodium carboxymethylcellulose-gelatin complex coacervation is fundamentally the same as the gelatin–acacia system.

   Microencapsulation of hydrophobic oils employing gelatin, carboxy-

methylcellulose and a second anionic colloid was accomplished by first preparing an emulsion at which coacervation does not occur. The mixture is
then acidified to promote coacervation and formation of microcapsules.
After chilling, the solid walls are treated with a cross-linking agent
(North, 1989).

Gelatin–cellulose acetate phthalate

Kassem and coworkers (1975b) investigated the coacervation of gelatin and
cellulose acetate phthalate and found that the optimal pH for the process
was 4.6. Polymer–polymer interactions were more important in dilute solutions just below the isoelectric point of gelatin and gave stable salt bonds.
Particle size increased with increasing concentration and decreasing temperature and also depended upon the rate of stirring.

Gelatin–chondroitin

Coacervates prepared from gelatin obtained from denatured tropocollagen
and chondroitin sulfate were investigated by Nagura *et al.* (1988). The coacervates were formed at pH 4.5 with a weight ratio of chondroitin sulfate
equal to 0.1. The helixes of the collagen molecules consisted of a small number of triple-helix crystallites. Intermolecular hydrogen bonds occurred between the amide groups of collagen and the hydroxyl groups of the
chondroitin molecules in the outer surface of the coacervate.

Gelatin–gantrez

Mortada *et al.* (1987a) investigated a number of parameters affecting the
complex coacervation of Type A gelatin and Gantrez-AN (G) polymers.
Gantrez-AN 119 (mol. wt 250 000) and 149 (mol. wt 750 000) are
polyvinylmethylether–maleic anhydride polymers and are soluble in water
with hydrolysis of the anhydride groups. In order to prepare the complex
coacervate, the gelatin solution at 40°C was added to a solution of Gantrez
at 40°C and stirred for 20 min and then cooled with stirring. Formaldehyde
was used to effect denaturation and this product was flocculated with
various alcohols. The sediment volume and sediment weight of the coacervate were determined after centrifugation and drying. The maximum
coacervation was achieved when the pH of the gelatin solution was 6.8, at
which equivalence of oppositely charged molecules were present. Increasing
the molecular weight of Gantrez decreased the combination with gelatin.

This was attributed to the coiled structure of G149, which possesses fewer available carboxylic groups for the reaction with gelatin. The optimum combination ratio for G119–gelatin is 1:4 and for G149–gelatin is 2:3 with a total concentration of 2.5% w/v for both polymers. The optimum concentration for denaturation with formaldehyde was 18% w/v at 2 h. The order of flocculation of the formaldehyde treated microcapsules was isopropanol = n-propanol > ethanol > methanol.

In a subsequent paper, Mortada *et al.* (1987b) described the encapsulation of nitrofurantoin in gelatin Type A and G119 or G149. The drug was added to the Gantrez solution and encapsulation was carried out in a manner similar to that described above. The encapsulation process was reproducible and about 90% of the drug was recovered in the microcapsules. The drug content decreased as the core to coat ratio decreased. The microcapsules were free flowing and tablets could be easily obtained by direct compression. The release of the drug in phosphate buffer at pH 7.4 and 37°C was decreased by encapsulation by using the Gantrez with the higher molecular weight and using a smaller core to coat ratio. The release kinetics were treated on the basis of a matrix model and yielded a linear relationship between drug concentration and $t^{1/2}$, thus following a diffusion-controlled model. Release data from capsules and tablets prepared from microcapsules were also obtained.

In a third paper, Mortada *et al.* (1988) investigated the bioavailability of nitrofurantoin microcapsules with a core to coat ratio of 1:2. The encapsulated product provided a prolonged release compared with that of the control formulation.

Gelatin–gelatin

Veis (1970b) studied the complex formation of gelatins with different isoelectric points of pH 9 and 5 as a function of initial mixing concentration. At a temperature of 20°C, which is below the conformational transition temperature of approximately 25°C, the fraction of gelatin in the coacervation phase increases with increasing mixing concentration but at 30°C the fraction decreases with increasing initial mixing conditions.

A complex coacervate of two oppositely charged gelatins has been prepared by Burgess and Carless (1985). They noted the previous work on this coacervate by Veis and coworkers from 1960 to 1967. The optimal concentration occurred when equal volumes of 1% deionized solutes of Types A and B were mixed together at 45°C, with stirring for 1 h. Subsequently, the temperature was reduced to 25°C for 4 h, then a 16% formaldehyde solution was added to harden the walls, followed by cooling to 4–5°C. After centrifugation and decanting the product was washed with water and isopropanol. The predicted optimum pH was 5.4, the electrical equivalence point, where the two gelatins have an equal and opposite charge. At this pH the electrophoretic mobility of the gelatins was low and was probably insufficient to effect coacervation. If the ionic strength is lowered, the electrophoretic mobility increases appreciably, promoting gelatin–gelatin coacervation. However, coacervation was not evident at 40°C and it was necessary to decrease the temperature to obtain complex flocculation. By controlled slow cooling, a more ordered gelation occurred, promoting coacervates with liquid, rather than flocculated properties. Concentrations of gelatin higher than 2.5% caused self suppression of the coacervation phase separation, likely due to the neutralization of charges, to form a large stable gel network. It was found that the shape of the droplets depended upon the final temperature. Higher temperatures (30°C) produced ellipsoid droplets, while at 15°C aggregation occurred. The authors suggest that the morphology of the droplets at 25°C is a result of the viscosity of the coacervate phase. The stirring forces may or may not be balanced by stabilizing forces within the droplet. Slower stirring speeds resulted in an increase in the droplet size and the fraction of amorphous droplets. The drug naproxen was encapsulated at a drug to colloid ratio of 1:5 at temperatures ranging from 5 to 30°C. The per cent drug encapsulated was highest at 25°C and the drug content of the microcapsule was higher when the microcapsules were produced at 30°C; however, the microcapsule yield was highest at 10°C.

Gelatin–gellan

Procedures for the microencapsulation of oils and solid particles using gelatin–gellan mixtures by complex coacervation were observed by Chilvers and Morris (1987) at low total polymer concentration and were limited to the pH range of 3.5–5.0.

Gelatin–genipin

Microcapsules with a high melting point have been prepared with gelatin and genipin for use in the pharmaceutical and food industries. For example, peppermint oil was encapsulated after emulsification with gelatin acidified with acetic acid and then treated with genipin and warmed to 40°C, washed and centrifuged (Kyogoku *et al.*, 1988).

Gelatin–inorganic

Coacervates have been prepared from a suspension of bentonite in alcohol
3:10 and a 10% solution of gelatin in water in a ratio of 20:1 by adding
the bentonite suspension to the gelatin solution with stirring. After filtration the coacervate was dried at 40°C. The granules were used with other
ingredients to prepare tablets for the protection of gastric mucosa (Oita
et al., 1982). Lenk and Thies (1986) have investigated the behaviour of acid
precursor gelatin with a polyphosphate in regard to pH, gelatin–phosphate
ratio and bloom strength. It was shown that the system exhibits classical
coacervation complex behaviour.

Gelatin–pectin

Microglobule size, morphology and recovery of pectin–gelatin coacervates
were investigated by McMullen and coworkers (1982). Coacervates were
prepared by combining solutions of pectin and Type A gelatin in varying
ratios at 45°C with stirring and adjusting the pH with NaOH solution.
After 2 min the pH was lowered with 0.5 N HCl and after stirring for
30 min, 5 ml of 37% formaldehyde solution were added. After cooling and
decanting, the microcapsules were suspended in glycerin and then treated
with an alcohol as the flocculating agent. The flocculated microglobules
were filtered and washed with isopropyl alcohol and dried. At a coacervation pH of 3.8 the mean globule size increases from 2 to 10 μm when the
pH of mixing was increased from 7 to 10. For solutions with equal pectin
and gelatin concentrations, the maximum yield of the coacervate occurred
at a colloid concentration of 2%. The maximum yield and microglobule
diameter occurred at a pH of about 3.8 after coacervation, but depended
upon the pH of mixing which ranged between 8 and 10. These changes were
related to the ionization of gelatin and pectin and the viscosity of the
microglobules. Increasing concentration of glycerin from 0 to 72% changed
the morphology from spheres to ellipsoids. The formation of ellipsoids was
attributed to dehydration of the coacervate and increasing intermolecular
association as a result of decreasing dielectric constant. Isopropanol and
1-propanolol produced satisfactory microglobules, while other alcohols
were not suitable.

In a subsequent paper, McMullen *et al.* (1984) encapsulated sulfamerazine with a gelatin–pectin coacervate. It was found that the drug should
be added at the starting pH, that is, before coacervation takes place. The
authors suggest that the drug is entrapped and the process is not a surface-
active phenomenon as suggested for gelatin–acacia. Globules of various

sizes with mean diameters 5.7, 9.2 and 25.5 μm containing 37–45% drug could be produced. The spherical shape of the microcapsules was maintained at drug loadings of ≤69% and ≤45% for 25 μm and 10 μm microglobules, respectively. A small suppression of coacervate yield occurred as the drug to colloid ratio increased, which was attributed to salt suppression by the drug. Complete digestion of the microglobules was observed with gastric and intestinal juice only. No apparent morphological change in the microcapsule was observed by extraction with 0.1 M HCl or 0.1 M NaOH or water. Several other drugs such as phenobarbital, hydrocortisone acetate and cod liver oil which have low solubility in water and small particle size were successfully encapsulated.

In a further paper, Bechard and McMullen (1986) investigated the dissolution times of gelatin–pectin microglobules as a function of formaldehyde concentration and reaction times. It was found that the dissolution half-lives, in terms of the number of microglobules, can be controlled over a period from 2.7 to 751 min in a solution of sodium chloride and polyoxyethylene sorbitan monolaurate. A decrease in dissolution rate of the microcapsules was observed with aging of the product stored at ambient conditions.

The gelatin–pectin coacervate has also been used to encapsulate indomethacin. Ku and Chin (1989) found that the optimum pH and pectin:gelatin ratio for microcapsules was 3.8 and 1.2 respectively. As the concentration of colloid solution increased, the wall thickness increased. The 50% release time for indomethacin prepared from 1, 1.5 and 2% colloid solutions were 3, 5 and 6 min, respectively, while that of indomethacin powder was 50 min.

Gelatin–polyvinyl alcohol

Cho et al. (1982) investigated the coacervation of gelatin and poly-(vinyl alcohol). The coacervation pH phase diagram showed a coacervate region consisting of two liquid phases and a non-coacervate region consisting of a single liquid phase. The intensity of coacervation was greatest at pH 5 and increased with increasing temperature. The coacervation was attributed to hydrogen bonding between the two polymers.

Histone–acacia

Coacervation of acacia has been studied by the nephelometric technique and the coacervate drops ranged in size from 0.5 to 500 μm. The average size and the number of drops served as parameters of the coacervate behaviour (Gladilin et al., 1972).

Sulfated poly(vinyl alcohol)–aminoacetalysed poly(vinyl alcohol)

Nakajima and Sato (1972) have discussed the complex coacervation of sulfated poly(vinyl alcohol) and an aminoacetalysed poly(vinyl alcohol). The three component system of the polymer salt, water and sodium bromide was investigated. The results are interpreted according to a theoretical equation for the free energy of mixing, taking into account the entropy and enthalpy contributions.

In a second paper, Sato and Nakajima (1974a) reported on the conditions for complex coacervation of the two polymers by relating the effects of charge density on phase separation. Conditions for coacervation were discussed as a function of chain length, interaction between polymer and water, the temperature, the electrostatic interaction and the number of charges on the polyelectrolyte chain. Subsequently, the authors, Sato and Nakajima (1974b,c), discussed the conditions for the formation of coacervate droplets as a function of charge density and polymer concentration. Furthermore, they indicated that the concentration of the coacervate phase at 25°C decreased and the concentration of the equilibrium liquid phase increased with increasing polymer concentration. The reduced viscosity of aqueous solutions of both polymers increased with decreasing polymer concentrations. The volume and polymer fraction of the coacervate phase containing the two polymers passed through a maximum value with increasing polymer concentration.

Okihana and Nakajima (1976) found that a 1:1 complex formed upon mixing the ratio of two polymers. The concentration of polymers in the coacervate and equilibrium liquid depended upon the initial polymer concentration. Coacervation of the 1:1 complex was suppressed by the addition of salts owing to a change in chain conformation.

COACERVATION-PHASE SEPARATION USING A SINGLE WALL-FORMING POLYMER SOLUBLE IN AN ORGANIC LIQUID

Acrylates

Hydrophobic compounds have been encapsulated with 2-diethylaminoethyl methacrylate–methacrylic acid-styrene copolymer latex by Ushiyama (1979). For example, castor oil was emulsified in water containing an anionic surfactant, Emal A, to a particle size of 30–50 μm. The pH was adjusted to about 9 and the polymer added, which has an isoelectric point of 7.2, and the pH adjusted to 9–10 with NaOH. After acidifying to a pH of 5–5.5 to form the capsules, the product was spray dried.

Donbrow *et al.* (1984) encapsulated potassium dichromate and paraceta-mol with poly(methyl ethyl methacrylate) (Eudragit Retard). The polymer and polyisobutylene were dissolved in chloroform and the core material was suspended in the solution. Cyclohexane containing polyisobutylene was added at a controlled rate and coacervate droplets formed which encap-sulated the core material. A decrease in the rate of addition of the non-solvent caused a decrease in the rate of release of the core material; this was attributed to structural changes as the core concentration was almost constant at about 80%. At high addition rates of the non-solvent, the microcapsules had polymer spheres attached to the surface – thus the effec-tive wall thickness was reduced and the release rate increased. As more non-solvent was added, more polymer came out of solution – thus the percen-tage of core material in the microcapsules decreased and the release rate decreased. It was also found that smaller particles gave faster release rates.

Chun and Shin (1988) encapsulated aspirin with Eudragit RS polymer from a solution of chloroform with polyisobutylene dissolved in cyclohex-ane. The polyisobutylene functioned as a coacervation-inducing agent and gave smooth microcapsules with less aggregation. By increasing the propor-tion of the wall material, particle size and wall thickness, and the concentra-tion of paraffin wax in the cyclohexane as a sealant, a product with sustained release characteristics could be obtained. Release was independent of pH of the medium and the mechanism of drug release from both non-sealed and sealed microcapsules appeared to fit Higuchi matrix model kine-tics. The aspirin microcapsules were more stable than free drug in a solution of $NaHCO_3$.

Eudragit L 100, a copolymer of methacrylic acid and methylmethacry-late which is insoluble in acid but soluble in alkaline solution, was used to encapsulate aspirin (Okor, 1988). The drug and polymer were dissolved in 95% ethanol and the solution was evaporated to dryness. The product was crushed and passed through a sieve and the fraction between $710 \mu m$ and $500 \mu m$ was collected. Dissolution was retarded in acidic medium, but enhanced in neutral medium. The author suggests that drug–polymer attractions are possibly stronger than drug–drug attractions, thus partly accounting for the delayed release in the acid medium. In the alkaline med-ium the polymer is soluble and readily liberates the aspirin.

Okor (1989) prepared colloidal solutions of ethyl acrylate (trimethyl ammonium) ethyl acrylate chloride–methyl methacrylate copolymer using ethanol as a solvent and water as the non-solvent. Stability of the dispersion to electrolytes such as NaCl and Na_2SO_4 increased considerably with an increase in the polymer cation content. The polymer dispersions were most sensitive to Na_2SO_4 and least sensitive to NaCl. In 1990 Okor encapsulated the drug salicylic acid with acrylate–methacrylate copolymers. The drug and polymer were dissolved in ethanol and excess water, the non-solvent, was added in the presence of a flocculating agent, NaCl. The dried coacervates were compressed into tablets or placed into capsules. It was found that drug release rates decreased exponentially with increase in polymer concentration in the coacervate, but increased exponentially with an increase in polymer cation content at a constant polymer concentration of 20% w/w. The increase in release rate was associated with an increase in polymer 'swell-ability'. Drug release rates from tablets were retarded compared with those from capsules; this was believed to be due to poor disintegration of the tablets.

A coacervation technique using an acrylate–methylacrylate copolymer was used to form an aqueous based coating system consisting of the water-insoluble copolymer and sucrose in varying ratios to coat matrix cores by Okor *et al.* (1991). Drug release rates increased as the concentration of the sucrose increased in the film coating. Doubling the coating thickness from 75 μm introduced a lag time for release of the model drug salicylic acid from 0.5 to 2.5 h depending upon the amount of sucrose. Overall, however, the release rates were hardly affected by the coating thickness.

Aqueous dispersions prepared by coacervation of Eudragit RL 100 and RS 100 were prepared by Okor (1991). A lower viscosity and higher gel point was observed with Eudragit RL 100. This phenomenon was explained by the higher degree of mutual repulsion of the cationic charges in Eudragit RL 100 compared with Eudragit RS 100. The fluidity of aqueous dispersion of these two polymers suggests their use in film coating processes.

Eudragit RS 100 polymer dissolved in chloroform was used to coat zipeprol hydrochloride. Cyclohexane containing polyisobutylene effected coacervation. The mechanism of drug release from the microcapsules appeared to fit the Higuchi matrix kinetics. Plasma concentration time curves suggested that the microcapsules can be used as a sustained release product (Yong and Kim, 1988).

Ferrous fumarate and ferrous sulfate were encapsulated by evaporation and various other methods using different Eudragit polymers. *In vitro* dissolution studies indicate the release was linear, but there was an inflection point that separates the initial fast release from the later, slower phase. In some cases a biphasic pattern was noted for larger size microcapsules, whereas a monophasic pattern was observed with small microcapsules. Particle size was the most important factor in determining the dissolution. The nature of the polymer and integrity of coating had a minor influence on dissolution (El Shibini *et al.*, 1989).

Kim *et al.* (1989) encapsulated a complex of dextromethorphan hydro-bromide and a strong cation-exchange resin with Eudragit RS by phase separation using a non-solvent. It was found that the release rate from the coated complex could be controlled by the amount of coating material. The effect of pH and the ionic strength on the release rate of the drug was also studied.

Sprockel and Price (1990) encapsulated the complex of chlorpheniramine maleate and a carboxylic acid cation-exchange resin. The complex was suspended in an acetone solution of polymethyl methacrylate, then emulsified in liquid paraffin containing various additives. After 12 h of stirring to permit the evaporation of the solvent, the microcapsules were collected, washed with hexane and dried. Several parameters and additives were tested and it was found that: (a) larger microcapsules were obtained if the concentration of the polymer was increased; (b) fine particles of bentonite, Veegum, carbon black or emulsion stabilizers, reduced the microcapsule size at 3% concentration, but increased the size at 6% owing to incorporation into the microcapsules; (c) silicone fluid 60 000 cp was more effective in reducing the microcapsule size than silicone fluid 50 cp; (d) magnesium stearate, glyceryl monostearate and stearyl alcohol reduced the microcapsule size; (e) formulations with higher coat to core ratios resulted in slower release of the drug from the microcapsules; (f) larger microcapsules released the drug at a slower rate than did smaller microcapsules.

Alex and Bodmeier (1990) encapsulated pseudoephedrine hydrochloride by preparing a solution of the drug in water and then preparing an emulsion in a solution of poly(methyl methacrylate) in methylene chloride with the use of a sonicator. This primary w/o emulsion was added to the external phase – water containing 0.25% poly(vinyl alcohol) as stabilizer – with stirring at 1500 r.p.m. in a small container with baffles for 10 min to give a w/o/w emulsion. The microcapsules were filtered and rinsed with water. Sonication resulted in the smallest droplet size and highest drug content. As the drug was not soluble in the polymer solution, it could not diffuse to the external aqueous solution. The method had good batch-to-batch reproducibility with respect to drug loading. The yield was above 95% and the particle size ranged from 50 to 500 μm. The drug content of the microspheres increased with drug loading, increasing amounts of solvent, polymer, and polymeric stabilizer. This last factor was attributed to an increase in the thickness of the adsorbed layer of the polymeric stabilizer and an increase in viscosity close to the droplet surface, resulting in a reduction in the rate of solvent and drug diffusion across the droplet interface into the continuous phase. The drug content decreased with increasing stirring time, increasing pH of the continuous phase and increasing volume of the internal and external aqueous phases.

Theophylline was encapsulated with Eudragit RS 100 using a solution of the polymer, and polyisobutylene in chloroform (Chattaraj *et al.*, 1991). Phase separation and rigidization of the deposited polymer was effected by using cold *n*-hexane. Polyisobutylene below 5.5% w/w did not produce uniform microcapsules, but aggregates. The drug content of the microcapsules was at maximum at 5.5% w/w of polyisobutylene at a fixed core to coat ratio. High percentages of polyisobutylene decreased the yield of the product. Dissolution at 37°C with increasing pH indicated that, as the core to coat ratio increased, the rate of dissolution also increased. As the percentage of polyisobutylene was increased in the preparation of microcapsules, the rate of release decreased. Bioavailability studies in rabbits indicated that prolonged release was obtained.

Badawi *et al.* (1991) encapsulated theophylline with Eudragit E and Eudragit L by non-solvent techniques. The best method to coacervate the drug is by using Eudragit E while the drug is dispersed in solution by the addition of a non-solvent. Eudragit E had a higher affinity for the drug and increased the surface drug by entrapment of the drug within the coat. Eudragit L formed a better barrier to the drug, but the microcapsules were less than satisfactory.

Cellulose acetate

Cellulose acetate was used as the wall material for preparing microcapsules of hydrocortisone by coacervation. The nearly spherical capsules in the range of 10–20 μm were formed and the release of the drug was sustained up to 7 days (Singh *et al.*, 1982).

Cellulose acetate butyrate

Sprockel and Prapaitrakul (1990) encapsulated paracetamol with the polymer cellulose acetate butyrate by employing three different emulsion techniques. In the emulsion solvent evaporation method (ESE) the drug was dispersed in the polymer solution using acetone as the solvent. This phase was then emulsified in a liquid paraffin solution containing 1% sorbitan monooleate and stirred at 1400 r.p.m. at room temperature until the solvent had evaporated. The microspheres were collected, washed with hexane and dried. In the modified emulsion solvent evaporation method (MESE) a limited amount of a non-solvent, hexane, was added slowly to the drug dispersion containing the polymer in acetone. This mixture was then emulsified as described above. The emulsion non-solvent addition method (ENSA) was the same as ESE, except that a limited quantity of solvent, hexane, was added to the emulsion after it had been stirred for 5 min. In the ESA method, the rate of solvent removal depends upon the rate of solvent partitioning into the mineral oil and solvent evaporation. In the MESE method, the addition of hexane likely increased the affinity of the external phase for acetone; as a result the rate of solvent removed depends primarily on acetone evaporation and this reduced the preparation time from 12 to 8 h. In the ENSA method, the removal of the solvent, acetone, depends primarily on the rate of solvent removal into the mineral oil and hexane solution and the preparation time was considerably reduced. The drug content and the drug release from the last two methods, MESE and ENSA, were significantly higher.

Cellulose acetate phthalate

A modified method based on the work of Kitajima *et al.* (1971) to prepare enteric coated microspheres using cellulose acetate phthalate was developed by Maharaj *et al.* (1984). Several pharmaceuticals such as loperamide and trifluoperazine-isopropamide and also rabies antigen were encapsulated by suspending the active compound, diluted with sucrose containing cornstarch if necessary, in paraffin oil. Then a solution of the polymer in acetone-ethanol 95% was added. Shortly thereafter the chloroform was added to harden the microspheres which were then decanted, washed and collected. The size of the microspheres increased as the time for formation increased from 0.5 min to 10 min and the encapsulation method had no appreciable effect on the activity of the biologically active substance.

In a subsequent paper, Beyger and Nairn (1986) prepared a three-component phase diagram of the system cellulose acetate phthalate, light mineral oil and solvent, acetone and 95% ethanol, to indicate the appropriate region for preparing microcapsules and also the effect of surfactant, sorbitan monooleate, concentration on the product. Chloroform was used as the hardening liquid. It was found that aggregation of the microcapsules could be minimized at low solvent concentration. In addition, pharmaceuticals could be microencapsulated regardless of their solubility in the polymer solvent or hardening liquid. The size of the product increased as the core to coat ratio was increased to a maximum (1.5:1). In addition, the time and order of addition of the various ingredients used to prepare the microcapsules was investigated. In general, the addition of the drug to the mineral oil followed by addition of the polymer solution was preferred, although other procedures gave satisfactory results. In some cases, however, it is necessary for the drug to be present as soon as the coacervate is formed for suitable microencapsulation. Particles with a large size, 20–50 mesh, were not always suitably coated.

Dibunol has been encapsulated with cellulose acetate phthalate. A 2% polymer concentration gave the most uniform product where approximately

91% of the microcapsules had a particle size of 160–250 μm (Berseneva
et al., 1988).

Mortada (1989) prepared microcapsules of phenobarbital by dissolving
both the drug and cellulose acetate phthalate in a 9:1 mixture of ethylacetate and isopropyl alcohol. Various factors affecting the coacervation
such as temperature, speed of agitation, polymer concentration and drug
content were studied. The release from the microcapsules was a function
of both particle size and core to wall ratio.

Ku and Kim (1989) encapsulated propranolol HCl with cellulose acetate
phthalate in a system containing paraffin, acetone and ethanol. The wall
thickness of the microcapsules increased with increasing cellulose acetate
phthalate concentration and the dissolution rate decreased. The dissolution
rate in both simulated gastric and intestinal fluid was determined.

Dharamadhikari and colleagues (1991) microencapsulated salbutamol
sulfate with cellulose acetate phthalate. The polymer was dissolved in
acetone and the drug was mixed with light liquid paraffin. Both phases were
mixed together and after evaporation of the acetone, the microcapsules
were collected and washed with ether and water. Free-flowing spherical
microcapsules were obtained. It was noted that the percentage of drug
encapsulated increased as the amount of coating polymer increased in the
coating phase. Microcapsules prepared with a coat to core ratio of 2:1
showed delayed release in an *in vitro* dissolution study during which the
pH was changed.

Terbutaline sulfate and propanolol hydrochloride were also encapsulated with cellulose acetate phthalate by Manekar *et al.* (1991, 1992), in
a similar manner as described immediately above. However, the dose of the
drugs are small, they were diluted with mannitol. Products with low coat
to core ratios were unable to prolong drug release when the pH was changed
from 1.2 to 7.5; however, microcapsules with a higher coat to core ratio
or with a mixture of the polymer and ethylcellulose provided release for
up to 12 h. Propanolol hydrochloride in cellulose acetate phthalate followed
a matrix mechanism of release in acidic media and a zero-order release
in alkaline media. When coated with a mixture of the two polymers, it
showed a zero-order release throughout the dissolution test as the pH was
changed.

Cellulose acetate trimellitate

Sanghvi and Nairn (1991, 1992) investigated the formation of microcapsules
using cellulose acetate trimellitate. Three-component phase diagrams were
prepared to show the region of microcapsule formation for the system

polymer, light mineral oil and the solvent acetone–ethanol. Chloroform was used as the hardening agent. Microcapsules were only formed when the polymer concentration was in the 0.5–1.5% range and the solvent concentration in the 5–10% range. The addition of surfactants such as sorbitan trioleate or sorbitan oleate to the mineral oil altered and/or increased the region of microencapsulation. Surfactants with higher hydrophile–lipophile balance values tended to decrease the area of microcapsules on the phase diagram. Sorbitan monooleate 1% in mineral oil gave products with smoother coats and a more uniform particle size. Tartrazine-containing microcapsules were prepared and the smallest microcapsule size was obtained when sorbitan monooleate 3% was used and these microcapsules had the slowest rate of release in an acidic medium. As a result of the removal of acetone from the polymer solution by the mineral oil, a polymer-rich phase is formed and after combining with other droplets and/or the core material, the microcapsules are formed which are hardened by further loss of solvent to the dispersion medium and also by the addition of chloroform.

Ethylcellulose

It was found that the time for release of sodium phenobarbitone from ethylcellulose microcapsules increased as the core:wall ratio decreased. With a constant core:wall ratio, the small microcapsules released their contents more rapidly than the larger ones (Jalsenjak *et al.*, 1976).

In a series of papers Donbrow and Benita (1977) investigated the effect of polyisobutylene on the coacervation of ethylcellulose. Ethylcellulose and polyisobutylene were dissolved in cyclohexane and the solution was allowed to cool slowly from 80°C to 25°C with controlled agitation. After 24 h, a clear upper phase containing the polyisobutylene and a lower phase of coacervate droplets formed whose particle size decreased with phase coacervation volume increase, which was increased by polyisobutylene. The product was a free-flowing powder, in contrast to the aggregated mass in the absence of polyisobutylene. The release of salicylamide from microcapsules showed first-order kinetics and the release rate increased with polyisobutylene concentration because of the thinner coating. It was indicated that polyisobutylene acts as a protective colloid in the process and prevents the agglomeration of ethylcellulose microcapsules.

Benita and Donbrow (1980), in a second paper, indicated that using a temperature reduction method for preparing coacervation droplets, in the absence of, or a low concentration of polyisobutylene, aggregates were formed, whereas higher concentrations of polyisobutylene stabilized the droplet. Polyisobutylene is not coprecipitated and acts as a stabilizer by adsorption. Increased concentration of polyisobutylene or higher molecular weights of polyisobutylene raised the phase coacervation volume and decreased the particle size indicating increased stabilization.

Benita and Donbrow (1982) employed polyisobutylene as a protective colloid to prepare microcapsules of salicylamide and theophylline based on the temperature differential solubility of ethylcellulose in cyclohexane. A minimum concentration of polyisobutylene was necessary to prevent aggregation and as its concentration was increased, it yielded microcapsules of higher drug content because the coating was thinner; furthermore, there was an increase in the release rate of the drug from the microcapsules. Microcapsule drug content decreased with decreasing particle size of the drug in the presence of the protective colloid. This was caused by a more complete uptake of the wall polymer on the increased surface of the core material.

A mixture of ethylcelluloses with a viscosity of 100 cp (0.1 Pa s) and a viscosity of 45 cp (0.045 Pa s) was used to encapsulate trimethoquinol using polyisobutylene as an agent to induce phase separation. The mixture was cooled from 78°C to room temperature and the microcapsules were filtered, washed and dried (Samejima and Hirata, 1979).

Samejima et al. (1982) prepared microcapsules of ascorbic acid with ethylcellulose using the temperature change technique. They found that polyisobutylene was better than either butyl rubber or polyethylene. The polyisobutylene changed the gel into a coacervate with the formation of smooth microcapsules with thick walls. The microcapsules did not aggregate appreciably and gave a slow release of the vitamin.

In a subsequent paper, Koida et al. (1983) used a similar method to encapsulate ascorbic acid with ethylcellulose using polyisobutylene. It was found that aggregation decreased with increasing molecular weight of ethylcellulose. The molecular weight of ethylcellulose which gave a minimum release rate was affected by the molecular weight of polyisobutylene. Polyisobutylene of high molecular weight gave less aggregation than polyisobutylene of low molecular weight. The relationship between the release rate and the molecular weight of ethylcellulose used depended primarily on the compactness of the wall, rather than its thickness.

In a patent, Samejima et al. (1984) described the encapsulation of trimebutine maleate with ethylcellulose using liquid paraffin and polyisobutylene in cyclohexane to give a solubility parameter of 7–10 $(cal\,cm^{-3})^{1/2}$, and then subsequent cooling. The product was free-flowing microcapsules.

Koida et al. (1984) investigated the effect of molecular weight of polyisobutylene on the microencapsulation of ascorbic acid using temperature reduction with a solution of ethylcellulose. After fractionating polyisobutylene, several fractions of various molecular weights were obtained. It was found that aggregation of the microcapsules decreased with increasing $\bar{M}$ (viscosity–average molecular weight), and above a value of 6×10^5 it was almost wholly prevented. The influence of $\bar{M}$ of polyisobutylene on the coacervation process was determined by measuring the volume fraction, the ethylcellulose content and the viscosity. It was found that the wall-forming temperature was lower with higher $\bar{M}$ of polyisobutylene. With higher $\bar{M}$ of polyisobutylene, a larger coacervation volume was produced, but the concentration of ethylcellulose in the coacervation phase was less and there was a very low concentration of polyisobutylene in the coacervate phase. The viscosity of the coacervate phase was higher with the lower $\bar{M}$ of polyisobutylene; this was attributed to the higher concentration of ethylcellulose in the coacervate. It was found that the temperature of the viscosity maximum coincided with the wall-forming temperature which appeared to be the most important temperature for microencapsulation. As the temperature decreases and reaches the temperature of maximum viscosity, the size of the ethylcellulose droplets gets larger and these gel-like droplets deposit on the surface of the drug and, after fusing, they form the wall. The effect of mixing high and low $\bar{M}$ polyisobutylene showed that with an increase of low $\bar{M}$ polyisobutylene, average wall thickness and compactness increases and the wall becomes less uniform.

Several different techniques have been employed to encapsulate ion-exchange resin beads containing benzoate with ethylcellulose by temperature change and non-solvent addition (Motycka and Nairn, 1979). Different viscosity grades of ethylcellulose, either alone or in conjunction with various plasticizing agents such as castor oil, butyl stearate and the protective colloid polyethylene were used. Some of these products were then treated with paraffin. In addition, the benzoate complex was encapsulated using gelatin and acacia and also cellulose acetate butyrate. It was found that the rate of release, as described by Boyd *et al.* (1947), could be controlled by the type of encapsulating material used and the phase separation process. The slowest rate of release was achieved with the microcapsules which were subsequently treated with paraffin. It was found that tough, dense films of large molecular weight compounds delayed the release of the anion. The decrease in the diffusion of the benzoate ions corresponded with an increase of the density of the film. Additives with the greatest lipophilic characteristics, polyethylene and paraffin produced the greatest resistance to ion transfer.

In a subsequent paper, Motycka *et al.* (1985) encapsulated ion-exchange resin beads containing theophylline with ethylcellulose, inducing phase separation by temperature reduction and by evaporation. Some of the products were subsequently treated with a solution of hard paraffin. Several products encapsulated with ethylcellulose by evaporation and also subsequently treated with a solution of hard paraffin gave a product that released the drug according to zero-order kinetics. It was found that the pattern and the rates of release could be controlled by the cross-linking of the resin and the coating procedure used.

Ethylcellulose microcapsules of ion-exchange resins containing theophylline were prepared by the evaporation method using ethylcellulose dissolved in ethyl acetate as the coating polymer, polyisobutylene dissolved in cyclohexane as a protective colloid and light liquid paraffin as the suspending medium (Moldenhauer and Nairn, 1990). Predominantly mononucleated microcapsules were formed by controlling the amount of ethylcellulose used, the particle size and the appropiate concentration of the protective colloid. The rate of release of the drug was altered by the cross-linking of the ion-exchange resin, the amount of ethylcellulose and the smoothness of the coat on the resin beads. Release rates from coated resin beads with low cross-linking followed a logarithmic plot indicating membrane controlled release, whereas coated resins with a higher degree of cross-linking followed a $t^{1/2}$ plot, indicating particle diffusion control.

In a subsequent paper, Moldenhauer and Nairn (1991) investigated the effect of the rate of evaporation on the coat structure of the microcapsules which were predominantly mononucleated. The rate of solvent evaporation influenced the surface morphology, the shape, and the porosity and the purity of the ethylcellulose coat. Microcapsules had tails and porous coats at slow evaporation rates. Faster evaporation rates resulted in the formation of microcapsules with no tails and smooth, but wrinkled coats. Coat porosity was minimal at intermediate evaporation rates. Microcapsules which showed rapid release rates of theophylline were formed when the very fast, slow and very slow evaporation rates were used to form the microcapsules. Intermediate evaporation rates formed coats with minimum porosity, leading to slow release rates of the drug.

Baichwal and Abraham (1980) encapsulated metronidazole by using ethylcellulose and polyethylene glycol 4000 in different proportions. As a result of encapsulation, the release of the drug was delayed and the percentage drug release, as a function of time, increased with increasing content of the polyethylene glycol.

Ascorbic acid has also been encapsulated using a solution of ethylcellulose in cyclohexane. The product had 2–3% wall material and a wall thickness of 6–10 μm (Shopova and Tomova, 1982).

Adriamycin was encapsulated with ethylcellulose in cyclohexane using the temperature reduction method (Kawashima et al., 1984). Polyisobutylene, rather than polyethylene, was found to be an effective coacervate-inducing agent. With increasing concentration of polyisobutylene, the average diameter of the particles decreased owing to reduced agglomeration. Microcapsules of the drug encapsulated with ethylcellulose at 2% polyisobutylene effectively prolonged the release of the drug compared with 1% or 3% polyisobutylene. The increase in rate of release noted when 3% polyisobutylene was used was attributed to a thinner wall. Kinetics of release of microcapsules prepared with 2% polyisobutylene were linear when plotted against $t^{1/2}$ suggesting a matrix type of release.

Using a non-solvent which resulted in the formation of an emulsion, Kaeser-Liard *et al.* (1984) encapsulated phenylpropanolamine hydrochloride with ethylcellulose. The drug, 95% of the particles $<40\,\mu$m, was suspended in a solution of ethylcellulose dissolved in acetone. With stirring, a solution of equal volumes of mineral oil and petroleum ether, the nonsolvent, were added over 90 min. During this period, the first emulsion of non-solvent in the polymer solution inverted to an emulsion of the polymer solution in the non-solvent at the same time phase separation took place. The microcapsules were then hardened with the addition of hexane at $-20°$C. After stirring in the cold, the microcapsules were filtered and dried. The microcapsules had a particle size in the 150–300 μm range and the yield was 90–100%. Several parameters were investigated, namely the volume of the non-solvent, the volume of the solidifying agent, rate of addition of the non-solvent, stirring rate, temperatures of the coacervation step and the hardening step and the core to wall ratio. The rate of drug release increased as the volume of the non-solvent was increased from 300 to 400 ml, as the temperature of hardening was increased from $-10°$C to room temperature, and as the core to wall ratio was changed. The rate of addition of the nonsolvent and the stirring speed did not affect the drug release from the microcapsules.

Sulfamethoxazole was encapsulated with ethylcellulose using an emulsion technique by Chowdary and Rao (1984). The drug was dispersed in a solution of ethylcellulose in acetone. This dispersion was added in a thin stream to stirred liquid paraffin which formed an emulsion. Water, the nonsolvent, was then added to cause coacervation and production of the microcapsules. After centrifugation, the product was washed with petroleum ether and then dried. Batches of microcapsules were prepared using different core to coat ratios. The time for 50% of the drug to be released in an acidic and neutral medium increased as the particle size increased and as the percentage of the coat material increased.

Chowdary and Rao (1985) described the influence of Span 60 and Span 80 on the preparation of microcapsules by emulsification. It was found that the inclusion of surfactants decreased the microcapsule size, but did not alter drug release. The drug release with or without a surfactant was similar for a particular size of microcapsule.

Chowdary and Annapurna (1989) encapsulated aspirin, metronidazole, paracetamol and tolbutamide by three different methods. Method I was coacervation-phase separation of ethylcellulose dissolved in toluene by the addition of petroleum ether. Method II was similar except that carbon tetrachloride was used as the solvent. Method III used thermal induction of the coacervate of ethylcellulose from cyclohexane. In all cases the drug was added to the polymer solution. The wall thickness was determined by the method of Luu *et al.* (1973). The apparent dissolution rate constants, K_{app}, were calculated from the initial slope of the release curve as described by Koida *et al.* (1986). The permeability constants, *Pm*, were determined from the following equation:

$$Pm = \frac{K_{app} VH}{ACs}$$

where V is the volume of the dissolution medium, H is the wall thickness of the microcapsules, A is the surface area of the microcapsules, and Cs is the solubility of the core in the dissolution medium.

The wall thickness ranged from 7.9 to 39.3 μm and the apparent dissolution rate constant ranged from 0.53 to 12.32 mg min^{-1}. It was found that for all four cores the order of permeability of the microcapsules was method III > method I > method II which suggests that the permeability depends upon the method employed.

Rak *et al.* (1984) prepared potassium chloride microcapsules using ethylcellulose by phase separation from cyclohexane by temperature change. It was noted that the addition of macrogol 300 or 4000 improved the formation of microcapsules and decreased the aggregation of the product.

Potassium chloride was encapsulated with ethylcellulose by coacervation with cyclohexane using polyethylene glycol by Chalabala (1984). The drug, with a particle size of 80 μm, had a microcapsule size of 125–187 μm with agglomerates up to 605 μm. High core to wall ratios gave smaller microcapsules.

Szretter and Zakrzewski (1984a) coated riboflavin with ethylcellulose dissolved in cyclohexane by the temperature change method. The solution also contained PEG 6000 and Tween 20. The product was stable at room temperature against oxidation, photodecomposition and humidity. The vitamin was also encapsulated with PEG 6000 by mixing at 70°C with paraffin oil, and ligroin containing PEG 6000 and Span 60 or Tegin G. The suspension was cooled to room temperature, filtered, washed and dried.

A mixture of ethylcellulose and polyethylene glycol 6000 has been used as a coating material and the process is carried out in cyclohexane to improve the stability of ascorbic acid (Szretter and Zakrzewski, 1987a).

Cisplatin was encapsulated with ethylcellulose dissolved in cyclohexane in the presence of low density polyethylene by the temperature reduction method (Hecquet *et al.*, 1984). Two stirring methods were used during the cooling stage – mechanical and sonication; however, no difference in microcapsule characteristics could be discerned between the methods. A number of different concentrations of drug, ethylcellulose and polyethylene were used to prepare the microcapsules and several observations were made: (a) losses of microencapsulated drug content occurred on increasing the ethylcellulose concentration; (b) the average drug content did not change if the amount of polyethylene was increased, but the proportion of small-size microcapsules increased; (c) the microcapsule composition appeared to be independent of particle size; (d) the wall thickness increased with an increase of ethylcellulose concentration. The drug was not decomposed by the microencapsulation process and certain products which released 80–100% of the drug within 24 h were selected for further studies.

Encapsulation of rifampicin was effected by dissolving ethylcellulose in ethyl acetate and adding the drug mixture. After stirring for 4 h petroleum ether was added at a controlled rate until coacervation started and then the mixture was stirred for 1 h. The microcapsules were collected, washed, dried and eventually made into pellet form (Khanna *et al.*, 1984).

Dihydralazine sulfate was encapsulated with ethylcellulose by Oner *et al.* (1984). The microcapsules were separated by size. The time for half of the drug to be released increased as the core to wall ratio decreased and as the particle size increased. Release appears to take place by diffusion.

Oner *et al.* (1988) encapsulated zinc sulfate using ethylcellulose dissolved in carbontetrachloride. Warm petroleum ether, a non-solvent, was added and the product was collected and washed with the non-solvent and dried. The rate of release in distilled water was determined and evaluated kinetically by the Rosin–Rammler–Sperling–Bennet–Weibull Distribution, which gave a good fit in defining the release from the microcapsules. A comparison of the release with hard gelatin capsules was also made.

Lin *et al.* (1985) encapsulated theophylline with ethylcellulose using four types of ethylene vinyl-acetate copolymer, with different concentrations of vinyl acetate (20–40%) as a coacervation-inducing agent. When *n*-hexane was added at the last step of microencapsulation, the particles aggregated except for the polymer containing 28% vinyl acetate. Using increasing concentrations of this polymer decreases the average diameter of the microcapsules as there was less aggregation. The wall thickness, the smoothness and compactness of the microcapsules increased and the porosity decreased with increasing concentration of the coacervating-inducing polymer. Differential scanning calorimetry indicated that the coacervate-inducing polymer was absent in all microcapsules.

Lin (1985) then investigated the influence of the coacervation-inducing agent ethylene vinyl acetate and polyisobutylene and cooling rates on the properties of microencapsulated bleomycin HCl. The particle size of microcapsules induced by ethylene vinyl acetate was smaller than that induced by polyisobutylene, and the size distribution of microcapsules using ethylene vinyl acetate depended on the cooling rate, which was different from that using polyisobutylene. The slower the cooling rate, the more prolonged was the release of the drug; this followed the Higuchi model. The time required for dissolution of 50% of the drug for both methods of microcapsule preparation decreased with an increase in the cooling rate. The rate-limiting step under certain circumstances was diffusion of the dissolution medium and the dissolved drug through ethylcellulose.

In a subsequent study, Lin and Yang (1986a) encapsulated chlorpromazine HCl with ethylcellulose using ethylene vinyl acetate copolymer as a coacervation-inducing agent. Higher concentrations of ethylene vinyl acetate decreased the microcapsule size and delayed the release of the drug because of the more compact surface and increased thickness of the wall. Microcapsules were compressed into tablets and prolonged the release considerably, which was attributed to a reduced surface area.

The release mechanism was discussed by Lin and Yang (1986b), and it was found that differential rate treatments showed that the release kinetics of theophylline from ethylene vinyl acetate copolymer-induced ethylcellulose microcapsules followed first-order kinetics.

Lin and Yang (1987) also encapsulated theophylline with ethylcellulose by temperature change using ethylene vinyl acetate copolymer as a coacervation-inducing agent. It was found that the higher the concentration of copolymer used, the more sustained was the release of the drug from the microcapsules. This was attributed to the lower porosity and thicker walls of the microcapsule. Bioavailability studies in rats indicated that microcapsules prepared with higher concentrations of ethylene vinyl acetate may act as sustained release forms.

Lin (1987) also investigated the effect of polyisobutylene of different molecular weights on the release behaviour of theophylline from microcapsules prepared with ethylcellulose. It was found that the release rate of the drug at pH values of 1.2 and 7.5 at 35°C was higher when polyisobutylenes with higher molecular weights were used. This was similar to results reported by Koida et $al.$ (1984). Several equations were investigated to study the release behaviour and one of the most useful was $1/y = A\ 1/x + B$ where y is the amount of drug released, x is the time, and A and B are constants that are proportional to the amount of drug released.

Cameroni $et\,al.$ (1985) encapsulated sulfadiazine by phase separation coacervation using temperature change and ethylcellulose and polyisobutylene was used as a protective colloid. Different release rates could be obtained by altering the wall thickness, which was controlled by the formulation. The rate of release for wall thickness $<5 \mu m$ followed the Hixson-Crowel theory and that for greater wall thickness followed the Higuchi theory.

Encapsulation of indomethacin and indomethacin modified by dry blending with a carboxyvinyl polymer by pulverization was carried out by Nakajima *et al.* (1987), with ethylcellulose and temperature reduction using polyethylene as a coacervation-inducing agent. The microcapsules were multinucleated and released the drug very slowly in a dissolution medium of pH 7.2. The rate of dissolution decreased as the amount of polyethylene was increased in the coacervation process. Subsequently, the microcapsules were prepared in the form of suppositories and were tested for dissolution characteristics.

Singh and Robinson (1988) investigated the effects of a number of surfactants on microencapsulation. Tweens and Spans, with HLB values ranging from 4.7 to 15, were used for the preparation of microcapsules of captopril. The process was carried out by dissolving the ethylcellulose in cyclohexane containing 2% absolute alcohol at 80°C. After dispersing the drug in this solution, it was cooled to room temperature and then to about 0°C. Microcapsules retained by 500–850 μm sieves were used for further studies. Dissolution tests in 0.1 N HCl at 37°C showed that the release rate decreased with an increase in the HLB of the surfactant. Based on the work of Barnett and Zisman (1959), who indicated that many solids will not be wetted if their critical surface tension is exceeded by the surface tension of the liquid, the authors suggested that the wetting for solvation of ethylcellulose with surfactants of higher HLB values resulted in an efficient coating around the drug particles and thus caused the slowest release. It was also found that higher ethylcellulose viscosity grades were less effective in extending the release of the drug in the concentrations used. This was attributed to the high viscosity of the coacervate droplets which inhibited coalescence and thus the formation of an intact ethylcellulose wall. Different kinetic models were used to explain the release. The best fit was the first-order kinetics plot with two straight lines that had two different slopes. The initial slope has a faster release than the terminal slope.

In a second paper, Singh and Robinson (1990) investigated the encapsulation of captopril using four viscosity grades of ethylcellulose with core to wall ratios of 1:1, 1:2, 1:3 by temperature reduction in cyclohexane. Dissolution studies in acidic media showed that the release depended upon the core to wall ratio and the viscosity grade of ethylcellulose and probably on the viscosity of the coacervate. A core to coat ratio of 1:1 showed that an increase of viscosity of wall material decreases the release rates. Viscosity grade 300 cp was not satisfactory for microencapsulation. The surface, as studied by scanning electron microscopy, showed that microcapsules prepared with 10 cp ethylcellulose were more porous and with larger pores than those prepared with 50 cp. The microcapsules did not fragment, alter shape or size or show enlargement of pores during dissolution. The *in vitro* release correlated better with biphasic first-order kinetics, rather than zero order or square root of time.

Singla and Nagrath (1988) encapsulated ascorbic acid to improve its stability in the presence of zinc sulfate. The microcapsules were prepared using ethylcellulose and the temperature change method using toluene as a solvent. The microencapsulated ascorbic acid was washed with toluene and dried in a vacuum. Several formulations including the product just described, ascorbic acid embedded in PEG 6000 or in stearic acid, were prepared in the form of tablets along with zinc sulfate. Tablets prepared from either the microcapsules or the stearic acid product had the maximum stability.

Based on a factorial design, the parameters which influence the particle size and particle size distribution of acetylsalicylic acid microcapsules coated with ethylcellulose were determined (Devay and Racz, 1988). The microcapsules were prepared by dissolving the drug and the polymer in varying ratios in diethylether in a reflux apparatus with stirring at 30°C. The solution was placed under vacuum and upon boiling, *n*-hexane was added slowly and the temperature reduced to 20°C; after filtration the microcapsules were dried. It was noted that the drug precipitated first and then became coated with the polymer. Coacervation was attributed to evaporation of the solvent, addition of a non-solvent and cooling. The parameters affecting the particle size for 50% through fall are in the order, rate of addition of hexane > drug content > ethylcellulose viscosity > speed of agitation. It was found that the standard deviation of the size increased with drug content, polymer viscosity, rate of addition of hexane and decreased with speed of addition.

Ferrous fumarate was encapsulated by phase separation using different ratios of ethylcellulose and castor oil. It was found that the drug release from the microcapsules depended upon the particle size, the thickness of the coat and the core:coat ratio (Shekerdzhiiski *et al.*, 1988).

Metoprolol tartrate was encapsulated with ethylcellulose using two different coacervation techniques by Nasa and Yadav (1989). In the non-solvent method, ethylcellulose was dissolved in a solution of carbon tetrachloride and the drug was added, then petroleum ether and talc. After decanting, the microcapsules were filtered and washed with petroleum ether. The second method involved temperature change of a solution of the polymer in cyclohexane. The product was filtered and washed with hexane and dried. It was found that the dissolution rate in distilled water of the product

prepared by the use of the non-solvent gave a slower dissolution. Furthermore, as the concentration of ethylcellulose used in preparing the microcapsules increased, the dissolution rate decreased. Stability studies of both pure
drug and microencapsulated drug showed similar results.

The effect of hydroxypropyl methylcellulose as a nucleating agent was
investigated using ethylcellulose and temperature change to effect microencapsulation. The core contained ascorbic acid, PEG 4000 and the nucleating
agent. Optimum conditions for the formation of microcapsules, such as
cooling, temperature, time, and concentration of hydroxypropyl methylcellulose, were assessed (Kaltsatos *et al.*, 1989).

Safwat and El Shanawany (1989) treated theophylline and oxyphenbutazone with a carboxyvinyl polymer, Carbopol CV 940, by dry blending
to control their release. The coated drugs were encapsulated with ethylcellulose using polyethylene and temperature reduction. The release rates of
the two drugs decreased as the content of polyethylene, a coacervation-
inducing agent, was increased, except at a concentration of 1% with the
drug oxyphenbutazone. Suppositories containing the microencapsulated
carboxyvinyl polymer modified drugs showed a pseudo zero-order release
profile. It was felt that this method, that is, surface treatment and microencapsulation, is a good one to prepare sustained release suppositories containing these drugs.

Vitamin C was encapsulated with ethylcellulose in cyclohexane using
ethylene polymer as the coacervation-inducing agent. It was found that the
dissolution rate of the microencapsulated vitamin and tablets was slower
than unencapsulated samples (He and Hou, 1989).

Shin and Koh (1989) investigated the effect of polyisobutylene on the
preparation of methyldopa encapsulated with ethylcellulose dissolved in
cyclohexane using temperature change. When polyisobutylene was used,
there was low aggregation and the surface of the product was smooth and
had a few pores. The dissolution of the drug was altered by the core to wall
ratio. The microcapsules were also treated with spermaceti, which reduced
the rate of release of the drug; the release was also influenced by the amount
of sealant used and the particle size of the product.

Chemtob *et al.* (1989) investigated the influence of polyisobutylene on
the microencapsulation of metronidazole by dissolving ethylcellulose in
cyclohexane at 80°C and cooling. The molecular weights of polyisobutylene
used were 3.8×10^5 and 1.12×10^6. As the concentration of polyisobutylene is increased, aggregation is minimized and spherical microcapsules are
obtained. At high concentrations of polyisobutylene some empty microcapsules are formed; this was also noted by Benita and Donbrow (1982). At a
concentration of 3%, the higher molecular weight polyisobutylene gave less
aggregation, similar to that reported by Koida *et al.* (1983). The percentage

of the sieve fraction $< 315\ \mu$m is generally increased when polyisobutylene is added during the preparation of the microcapsules. It was found that the total drug content was not generally influenced by the addition of the polyisobutylene. It was noted that the times for 50% of the drug to be released at a pH of 1.2 and 37°C decreased as the concentration of polyisobutylene with the lower molecular weight increases at a core to wall ratio of 1 to 1, but not at a ratio of 2 to 1. When polyisobutylene with a higher molecular weight was used T50% varied with polyisobutylene concentration.

Piroxicam microcapsules were prepared by coacervation using ethylcellulose. It was found that it took 240 min to release 63% of the drug from microcapsules, compared with 6.9 min for the drug in hard gelatin capsules (Bergisadi and Gurvardar, 1989).

Dubernet et al. (1991) prepared microcapsules of ibuprofen with ethylcellulose dissolved in methylene chloride. Methylcellulose or polyvinyl alcohol, as the emulsifying agent, was dissolved in water and then the polymer solution containing the drug was added, an emulsion formed and evaporation proceeded until all the solvent was lost. In addition to the above procedure the crystal window concept was used in which the solvent evaporation is interrupted and the supernatant removed to prevent crystallization in the aqueous phase. This tends to remove drug molecules and prevent deposition. However, crystal formation was in some cases observed in both systems. Based on altering the drug concentration in both the aqueous and non-aqueous phases and the nature of the emulsifier, a mechanism for crystal deposition is proposed which involves the formation of nuclei in the unstirred layer surrounding the emulsified droplet during solvent evaporation. Crystal growth is also controlled by the drug concentration in both phases and the viscosity of the polymer layer at the interface.

Propranolol hydrochloride microcapsules were prepared by solvent evaporation by dissolving ethylcellulose in acetone and adding the dispersion to liquid paraffin by Ku and Kang (1991). The amounts of drug dissolved at pH 1.2 in aqueous solution increased as the drug content of the microcapsules increased and the dissolution was not affected by the concentration of sorbitan tristearate in the microencapsulation process.

Bacampicillin was encapsulated with different viscosity grades of ethylcellulose dissolved in cyclohexane, employing polyisobutylene with different molecular weights (Oppanol B200, B100, B50, B3) as the coacervation-inducing agent (Kristl et al., 1991). It was found that when polyisobutylene of low molecular weight was used agglomerates are formed as a result of large coacervate droplet size and low viscosity of the continuous phase. If a high molecular weight of polyisobutylene is used, much of the ethylcellulose was not used for wall formation. Further experiments were carried out with Oppanol 50, and different organic liquids were used for washing purposes. It was found that a non-agglomerated, free-flowing product was obtained when *n*-heptane was used, in contrast to some agglomeration obtained when petroleum ether or cyclohexane were used. Different celluloses and different core to wall ratios influence the shape of the microcapsules. Usually, spherical and small microcapsules were obtained using ethylcellulose N-50 with a core to wall ratio of 1:1.5. Stability studies showed that most of the original drug was retained. A kinetic analysis of the release of the drug was carried out. It was found that a combined zero- and first-order kinetic relationship was most suitable. The drug release decreased with increasing molecular weight to a minimum when the molecular weight of ethylcellulose was approximately 13×10^4, depending upon the polyisobutylene used, then the rate of release increased with increasing molecular weight of ethylcellulose.

Chloroquine phosphate and quinine hydrochloride microcapsules have been prepared by a thermally induced coacervation method using ethylcellulose. The microencapsulation process masked the taste of the drug and dissolution studies showed a prolonged release profile. Tablets of the microencapsulated drug were also prepared and tested (Chukwu *et al.*, 1991).

Sveinsson and Kristmundsdottir (1992) encapsulated naproxen by coacervation-phase separation from a warm solution of ethylcellulose. The product after cooling was washed with cyclohexane and dried. The core to wall ratio was 1:1 or 1:2 and polyisobutylene concentrations ranged from 0 to 8%. It was found that an increase in the speed of stirring produced a greater proportion of smaller microcapsules, but dissolution characteristics and drug loading remained unaffected. Results of the sieving analysis indicated that the presence of polyisobutylene resulted in a pronounced decrease in the size of the microcapsules at both core to wall ratios. On increasing the concentration of polyisobutylene, the surface of the microcapsule became smooth and compact, but the shape remained irregular. The microcapsules were composed of aggregates of individually coated particles. The time for 50% of the drug to be released at a pH of 7.5 decreased from 140 min for 0% polyisobutylene to 20 min for 6% polyisobutylene when the core to wall ratio was 2:1.

Indomethacin was encapsulated with ethylcellulose by complex emulsification. By altering the core to coat ratio, the size range of microcapsules, or by incorporating a channelling agent such as PEG 4000 the drug release rate can be controlled (Jani *et al.*, 1992).

Puglisi *et al.* (1992) prepared microspheres of tolmetin by cooling a solution of ethylcellulose containing polyisobutylene or ethylene vinyl acetate copolymer. The presence of the coacervating agent did not appreciably influence the drug content or the wall thickness, but did increase the particle size, especially when polyisobutylene was added. Coacervation with either agent produced a smooth surface and fewer holes were observed with ethylene vinyl acetate, by both scanning electron microscopy and fluorescent microscopy. Dissolution studies were carried out at 37°C at pH 7.4 and 4 in aqueous medium and also in the presence of Tween 20. In all cases the encapsulated drug delayed the release. Gastric lesions produced by a tolmetin preparation in rabbits were reduced when the drug was encapsulated; this was attributed to a shorter contact time with the gastric mucosa. A decrease in body temperature effected by the drug and the encapsulated drug with or without a coacervating inducing agent was similar.

Hydroxypropyl methylcellulose phthalate

Morishita *et al.* (1973) encapsulated kitasamycin tartrate by emulsification and evaporation of 10% w/v solution of hydroxypropyl methylcellulose phthalate in a solution of acetone and methanol which was emulsified in paraffin oil at 5°C. The antibiotic, with a particle size of 200–500 μm, was then added with stirring and the emulsion was slowly heated to 30°C to give, after 3–4 h of evaporation, microcapsules with an enteric coat and a size of 300–700 μm.

   Encina *et al.* (1992) prepared a three-component phase diagram for the system hydroxypropyl methylcellulose phthalate, light mineral oil and acetone to show the region at which phase separation and microcapsules could be formed. The addition of small quantities of the surfactant sorbitan monooleate or sorbitan trioleate increased but sorbitan monolaurate decreased the region of the phase diagram where microcapsules were formed. Increasing the concentration of the surfactant did not affect the particle size appreciably, but an increase of polymer concentration increased the size of the microcapsules at all levels of surfactant concentration.

Poly β-hydroxybutyrate hydroxyvalerate

The preparation of reservoir type microcapsules was described by Embleton and Tighe (1992). The initial w/o emulsion was prepared by adding an aqueous gelatin solution to the polymer dissolved in dichloromethane, shaking and cooling and then transferring to a large volume of polyvinyl alcohol solution. Solvent evaporation occurred with stirring for 5 h. The microcapsules were sieved, washed with water and dried. A series of nine different poly β-hydroxybutyrate hydroxyvalerate polymers, in which both the molecular weight and hydroxyvalerate content were altered, were used. Microcapsules prepared from low molecular weight homopolymers were non-porous but shrivelied. These features disappeared as the molecular weight increased. Decreasing the molecular weight of the copolymer produced particles that were distorted and had a macroporous surface. Increasing the temperature to 40°C after phase combination usually produced a smooth, less porous particle.

Poly(lactic acid)

The rate of release of thioridazine from polylactide microcapsules, prepared by solvent evaporation from oil in water emulsions, was enhanced by the use of a base, NaOH. The rate of drug release depended upon the amount of base added to the aqueous phase of the emulsion. Using the results from scanning electron microscopy, it was suggested that the drug release could be due to modification of the internal structure of the microspheres during their preparation (Fong *et al.*, 1987).

Bodmeier and McGinity (1987c) encapsulated quinidine and quinidine sulfate with poly(DL-lactic acid) by the solvent evaporation method. The drug and the polymer were dissolved with heat in methylene chloride and this solution was then emulsified into the aqueous phase containing polysorbate 80 and at the pH of minimum drug solubility to minimize drug loss to the aqueous phase. Stirring was continuous until the organic solvent evaporated. The product was then filtered, washed with water and dried.

In a second paper dealing with quinidine and poly(DL-lactic-acid), Bodmeier and McGinity (1987b) showed that the drug loss to the aqueous phase occurred within the first 1–2 min of the emulsification step, as the pH was changed from 7 to 12 or 12 to 7. They suggested that the ability to change the pH without influence to the actual drug content within the microcapsule may permit the preparation of microcapsules at extended pH values. An increase in the volume of the aqueous phase resulted in an increase of drug content in the microcapsules. This was attributed to faster precipitation of the polymer at the droplet interface, as a result of polymer solvent diffusing into the water. An increase of temperature from 0 to 35°C during the formation of the microcapsules caused a decrease in the quinidine content of the product. This was attributed to an increase in the solubility of the drug in the aqueous phase. The higher temperature also caused an increase in the vapour pressure of the polymer solvent, leading to an increasing flow across the interface, resulting in film fracture.

In a subsequent paper, Bodmeier and McGinity (1988) reported on solvent selection for the preparation of microspheres by the evaporation method using poly(DL-lactide). The successful encapsulation of the drug within the microsphere was associated with: (a) a fast rate of precipitation of the polymer from the organic phase; (b) a low water solubility of the drug in the aqueous phase; and (c) a high concentration of the polymer in the organic phase. It was found that the rate of polymer precipitation was strongly influenced by the rate of diffusion of the organic solvent into the water phase. Organic solvents with low water solubility resulted in a slow polymer precipitation, permitting the drug to partition fully into the aqueous phase. Water-miscible organic solvents, when added to the organic phase, improved the drug content in the microspheres. The preparation of a solubility envelope for the polymer and an envelope for microsphere formation based on three-dimensional solubility parameters was useful for the selection of suitable solvent mixtures and the interpretation of solvent, non-solvent, polymer interactions and the formation of the microspheres.

Spores and viable cells were encapsulated with poly(lactic acid) dissolved in dichloromethane. Either spores or nutrient broth containing viable cells were added to the polymer solution. Then this suspension was added to a methylcellulose solution and the mixture stirred until the solvent evaporated. After filtration the product was washed with water and air dried. The core material was also encapsulated with gelatin and acacia using the complex coacervation method. The microcapsules produced were larger using the solvent evaporation method. Both methods permitted the encapsulated material to retain some viability. The solvent evaporation method was simple and more reproducible (Pepeljnjak, 1988).

In a series of papers Jalil and Nixon (1989) investigated the preparation and properties of microcapsules using poly(L-lactic acid) or poly(DL-lactic acid). In the first paper the phenobarbitone was microencapsulated by dissolving the polymer, poly(L-lactic acid), and drug in dichloromethane and dispersing the solution in 1% aqueous gelatin solution, to give an o/w system. With subsequent evaporation, the drug was found to be poorly encapsulated and microcapsules were small. In the other method of preparation, w/o, a solution of drug and the polymer in acetonitrile dispersed in light liquid paraffin containing Span 40 was allowed to evaporate. Drug loading in this system was high and the large microcapsules had a more porous surface.

In their next paper, Jalil and Nixon (1990b) used the poly(DL-lactic acid), acetonitrile, light liquid paraffin system for preparing phenobarbitone microcapsules. With an increase in temperature for evaporation, the surface of the microcapsules became more irregular and porous owing to deposition of phenobarbitone near the surface of the microcapsules. As the polymer concentration was increased, the surface became more irregular and non-continuous owing to rapid precipitation of the polymer, and the microcapsules became larger. The encapsulation efficiency was not appreciably affected by changes in temperature of preparation and polymer concentration. When the initial core loading was decreased the encapsulated efficiency decreased.

Jalil and Nixon (1990c), again using poly(DL-lactic acid), found that as polymers with lower molecular weights were used the microcapsule size decreased and the rate of swelling in an aqueous environment was greater. The gross morphology, encapsulation efficiency and density were not affected by changes in the molecular weight. In subsequent papers Jalil and Nixon (1990d,e) investigated the effect of polymer molecular weight on release kinetics and storage on microcapsule characteristics.

Gentamycin sulfate was encapsulated with poly(L-lactic acid) by adding the drug to a solution of the polymer in methylene chloride (Sampath *et al.*, 1992). Coacervation was induced by adding hexane at a controlled rate. Hardening was achieved by stirring for 2 h and the product was washed with hexane and allowed to dry. After sieving, the 125–450 μm fraction was used for further studies. A volume-based size distribution indicated a mean diameter of 343 μm, and a mean diameter of 14.8 μm was obtained based on particle number. This discrepancy was explained in terms of the breakup of aggregates. Dissolution studies at pH 7.6 showed that microcapsules with higher drug loading released their contents faster, and complete release ranged from 3 days to 3 weeks. Cylindrical implants were prepared by compressing the microcapsules in a punch and die and several dissolution studies were made on these products.

Poly(lactic-co-glycolic acid)

Lewis and Tice (1984) encapsulated several steroids, namely norethisterone, norgestimate, testosterone propionate, oestradiol benzoate, progesterone and levonorgestrel, with poly(DL-lactide-co-glycolide) using a solvent evaporation method. The quality of the microcapsules was determined by scanning electron microscopy and rate of release.

Thyrotropin-releasing hormones or analogues have been encapsulated with lactic acid-glycolic acid copolymer by means of an emulsion (Heya *et al.*, 1988). The hormone was dissolved in water and this solution was added to a solution of the polymer in dichloromethane with stirring. This emulsion was cooled to 18°C and then poured into a solution of poly(vinyl alcohol) and stirred to give a water/oil/water emulsion. The internal water/oil emulsion was solidified by evaporating the solvent and then the microcapsules were collected and freeze dried. The efficiency of encapsulating the hormone was 95.9% and the product contained 7.5% of the hormone.

Poly(DL-lactic-co-glycolic acid) copolymer 50/50 has been used to encapsulate Triptoreline, an analogue of luteinizing hormone releasing-hormone by Ruiz et al. (1989). Phase diagrams were prepared by dissolving the polymer in methylene chloride and then adding successive portions of silicone oil of a specific viscosity from 20 to 12 500 cs (2×10^{-5} m^2 s^{-1} to 1.25×10^{-2} m^2 s^{-1}). It was observed from the phase diagram that there were four steps involved as the amount of phase inducer was increased. First, the silicone oil produced a pseudo emulsion in the organic phase, then the beginning of phase separation occurred consisting of unstable droplets, next were a stable dispersion of poly(DL-lactic-co-glycolic acid) droplets were formed (this is called the stability window) and finally at the highest concentration the silicone oil aggregation of the droplets took place leading to precipitation. It was found that increased amounts of the silicone oil were required to reach the stability window as the concentration of lactic acid in the polymer increased. No stability window could be observed in the phase diagram when silicone oil with a viscosity of 20 cs (2×10^{-5} m^2 s^{-1}) was used; however, as the viscosity of the oil increased to 12 500 cs (1.25×10^{-2} m^2 s^{-1}), the area of the stability window increased. It was found that the more hydrophobic the copolymer, the more methylene chloride is a good solvent and the more silicone oil is required to desolvate the polymer. Microcapsules of the drug were prepared by suspending it in the polymer solution and silicone oil was added to effect coacervation. A non-solvent was then added to harden the product. It was found that as the volume of the phase inducer increases, the microsphere average diameter increases and the calculated specific surface decreases and consequently the core loading increases and the initial burst effect of release lowers.

In a subsequent paper Ruiz et al. (1990) fractionated poly(DL-lactic acid-co-glycolic acid) copolymer 50/50 by exclusion chromatography to give five different batches and determined the size exclusion chromatographic data for the fractionated polymer such as the number average and the weight average molecular weights. It was concluded that the polymer solvent affinity is mainly modified by the variation of the average molecular weight owing to differences in solubility. The lower the average molecular weight, the better methylene chloride serves as a solvent for the polymer. For microencapsulation purposes, polymers with an intermediate molecular weight of 47 250 were more suitable in terms of core loading and release purposes.

Multiphase microspheres of water-soluble drugs such as chlorpheniramine maleate, procainamide hydrochloride, and promazine hydrochloride were prepared by Iwata and McGinity (1992). The drug was dissolved in a dilute solution of gelatin and Tween 80. This was added to a solution of aluminium monostearate in soybean oil containing Span 80 and stirred to form a coarse w/o emulsion and then micronized. The polymer poly(DL-lactic acid) or poly(DL-lactic-co-glycolic acid) was dissolved in acetonitrile. Then the w/o emulsion was poured into the polymer solution and dispersed to give a w/o 'w' emulsion. Finally, this w/o 'w' emulsion was poured through a narrow nozzle into mineral oil containing Span 80 and the mixture stirred for 24 h to evaporate the acetonitrile. The hardened microspheres were filtered using nylon screen and washed with hexane, a Tween 80 solution, then water and dried. Drug loading efficiencies of 80–100% were obtained under specific conditions. The drug loading efficiency in the microspheres depended upon the ratio of the water/oil emulsion to polymer, and the concentration of surfactant in the mineral oil. Compared with conventional microspheres in which fine particles are homogeneously dispersed in the polymer beads, the multiphase microspheres allow a higher efficiency of encapsulation of water soluble drugs and eliminate the partitioning into the polymer acetonitrile phase.

Polystyrene

Phase diagrams for the system polystyrene-benzene-butanol were prepared by Bardet *et al.* (1969). The area of coacervation was defined and the different phases characterized in terms of their composition. The process of coacervation was related to the insolubilization of the polymer as a result of the strong interaction between the solvent and the non-solvent.

The coacervation of polystryene in a solution of cyclohexane by temperature lowering was investigated by Iso *et al.* (1985b). The polydispersity of the polymers was the important factor either to define the separation temperature for the separation of droplets or to determine the equilibrium composition of the dilute and coacervate phases. Microcapsules of glass beads had a thin film of the polymer and a thick coat of talc. The low efficiency of polymer utilization was improved by using a non-solvent for the polymer. Sodium sulfate crystals were encapsulated and the dissolution was related to the Higuchi model. It was found that the effective diffusion coefficient decreased as the encapsulation temperature was lowered.

In another paper, Iso *et al.* (1985a) investigated the three-component system polystyrene, cyclohexane and hexane and encapsulated glass beads and anhydrous sodium sulfate. The control of wall thickness was easier than when cyclohexane was used alone. The wall thickness was effectively controlled by adjusting the polymer concentration, the temperature of encapsulation and the amount of hexane added. The dissolution followed the Higuchi model.

Polystyrene was used to encapsulate frusemide or frusemide–PEG 6000 solid dispersion by El Shattawy *et al.* (1991). Coacervation was achieved by preparing a suspension of the drug in a solution of polystyrene in cyclohexane and effecting coacervation by adding the non-solvent petroleum ether. Dissolution studies *in vitro*, LD_{50} studies and oral toxicity of the polmyer were carried out in mice. The most suitable product was frusemide, PEG 6000, polystyrene with a weight ratio of $2:2:1$. The dissolution of this product was slower than that of the pure drug and faster than that of the pure drug in polystyrene microcapsules. The toxicity studies showed good agreement between the increase in LD_{50} and the decrease in dissolution rate.

Polyvinyl acetate

Microcapsules of phenylpropanolamine hydrochloride were prepared by adding a suspension of the drug in a solution of polyvinyl acetate, Rodopace, or polyvinyl acetate copolymer in acetone to petroleum ether, the non-solvent (El Shattawy *et al.*, 1992). The drug was also encapsulated using pan coating and an air suspension technique with a specified polymer solution or carnauba wax or hydrogenated castor oil solution. Several different formulations were tested. In general, microcapsules prepared by the coacervation phase separation method did not show suitable prolongation of release in dissolution studies in 0.1 N HCL compared with the other procedures. This was attributed to the porous coat leading to rapid leaching of the drug. Marked prolongation of dissolution was observed with the pan coating technique. This was attributed to the numerous thin coats that were applied and the large size of the microcapsules. It was found that the LD_{50} increased from $750 \, \text{mg kg}^{-1}$ for pure drug, to $1200 \, \text{mg kg}^{-1}$ for microcapsules prepared by air suspension, to $1500 \, \text{mg kg}^{-1}$ for microcapsules prepared by pan coating. The comparison of *in vitro* to the *in vivo* studies showed close agreement between the increase in lethal dose and the decrease in dissolution rate.

Polyvinyl chloride

Polyvinyl chloride microcapsules containing sulfamethoxazole were prepared by dispersing the polymer in *n*-hexane, effecting dissolution with chloroform and then subsequent coacervation with the non-solvent *n*-hexane. After cooling for a period of time at $5 \, ^\circ C$, the microcapsules were washed with cold non-solvent and dried. A three-component phase diagram was prepared to indicate the region of coacervation. Several parameters were investigated and it was shown that as the stirring speed increased, the sieve size of the microcapsules also increased. An analysis of the dissolution profile suggests that the drug release is controlled by diffusion, coupled with a dissolution process. Prolonged drug release was obtained over a period of 8 h (Das and Palchowdhury, 1989).

Pyridine polymers and copolymers

Several polymers, homopolymers of 4-vinyl pyridine, 2-vinyl-5-ethyl pyridine; copolymers of 4-vinyl pyridine with styrene, methyl acrylate and acrylonitrile; and 2-vinyl-5-ethyl pyridine with vinyl acetate, methyl acrylate and acrylonitrile were used to formulate chloroquinine phosphate for its taste abatement (Gupta and Agarwal, 1983). The drug was dispersed in a solution of the polymer in methanol. Then the non-solvent, ether, was added at a controlled rate. After cooling to harden the product, it was filtered and dried. Several products had satisfactory release in gastric fluid and at pH 7. Only four products, each with a different polymer, were formulated as dry suspensions. Aging tests were also carried out and dissolution characteristics were not affected by storing at 45°C for 2 months.

Rosin

Sheorey and Dorle (1990) described a method for encapsulating sulfadiazine with rosin using benzene as the solvent and this was emulsified with stirring into an aqueous bentonite suspension. After evaporation, the microcapsules were filtered, washed and dried. It was found that as the core to coat ratio increased from 1:1 to 4:1 , the drug in the microencapsulated product increased from 50 to 91% and the mean diameter decreased from 924 μm to 238 μm and the time for 50% of the drug to dissolve decreased from more than 180 min to 80 min in gastric fluid pH 1.2 and 50 min in intestinal fluid pH 7.6. It was found that bentonite, rather than ionic and non-ionic emulsifiers and other protective colloids such as gelatin or acacia, minimizes aggregration of microcapsules.

Sheorey and Dorle (1991a) again used the solvent evaporation method to encapsulate sulfadiazine with rosin. Ten different organic solvents with different rates of evaporation were used to dissolve the rosin, for example ether (fast), chloroform (medium) and petroleum ether (slow). Solvents with fast relative evaporation rates gave large microcapsules and lower drug content, and an increase in wall thickness and surfaces with many pores and fissures compared with solvents with slow relative evaporation rates. The authors suggest that quick evaporation of the polymer solvent causes rapid agglomeration of the visco-elastic polymer droplets and subsequent drying and rigidization occurs rapidly before a uniform coating of the drug occurs; this leaves a loosely deposited wall structure. Dissolution studies showed that while ether produced thick-walled microcapsules, the rate of release was rapid because of the surface characteristics. The influence of solid dispersing agents on the formation of rosin microcapsules was also investigated by Sheorey and Dorle (1991b).

Shellac

Sheorey *et al.* (1991) encapsulated sulfadiazine with shellac, using solvent evaporation. The drug was dispersed in a solution of shellac dissolved in isobutanol which has a higher boiling point than water. This suspension was then added slowly in a thin stream into a freshly prepared aqueous bentonite solution at 70°C. Phase separation occurred and the drug was coated upon evaporation of the solvent. The microcapsules were filtered, washed with water and dried. While a dry product could be obtained without the use of bentonite, the use of this agent promoted the formation of spherical microcapsules, whereas in its absence dry flakes and needles were obtained. No coalescence or aggregation occurred during the process and only a negligible amount of bentonite was incorporated into the microcapsules. Dissolution in water at pH 1.2 showed that as the core content increased, the microcapsules decreased in size and the rate of dissolution increased. This suggests a decrease in the wall thickness. At higher bentonite concentrations smaller microcapsules with a narrower size distribution were obtained; this was attributed to an increase in viscosity of the dispersion medium which reduced the aggregation of the globules. An increase in the rate of stirring from 1000 to 4500 r.p.m. reduced the mean diameter from 265 μm to 235 μm.

Various polymers

Three polymers have been used to encapsulate a hydrophobic drug (Seiyaku, 1980). For example, polystyrene was dissolved in an organic solvent which was immiscible with water. A hydrophobic drug was then incorporated into the solution and then this core material was dissolved in an enteric high-polymer electrolyte; for example, methylacrylate. Coacervation was then carried out with a solution of gelatin at an appropriate pH and cooling. Thus the drug is surrounded by the high-polymer and a coacervate wall film of gelatin and the polymer electrolyte.

Mathiowitz *et al.* (1988) prepared polyanhydrides of the following diacids: sebacic, bis (*p*-carboxy phenoxy) propane and dodecanedioic acid. The polymers were characterized by infrared spectroscopy, X-ray diffraction, viscosity, differential scanning calorimetry and scanning electron microscopy. Microspheres were prepared by dissolving the polymer in methylene chloride, adding the core and then the mixture was dropped into silicone oil containing Span 85 and varying amounts of methylene chloride, depending on the polymer. After stirring for 1 h, petroleum ether was added, and after further stirring the microcapsules were isolated. Several modifications of the above procedure were tested. In general, the surface of the microspheres was smooth with no pores. The porosity of the microcapsules depended upon the polymer and the proportions of the polymer used to prepare the microspheres. It was proposed that the process used with low molecular weight polymers took place slowly, resulting in relatively non-porous microspheres, whereas the process used for high molecular weight polymers was rapid and resulted in spheres with significant internal porosity. Drug release was affected by polymer composition, physical properties of the microspheres and the type of drug. Microspheres loaded with insulin showed good urine and serum glucose control in diabetic rats.

Bodmeier and Chen (1989) encapsulated three anti-inflammatory agents, namely indomethacin, ibuprofen and ketoprofen, by solvent evaporation using various polymers: ethylcellulose, poly(ϵ-caprolactone), poly(methyl methacrylate), polystyrene and Eudragit RS and RL. The polymer and drug were codissolved in a water-immiscible organic solvent. This solution was then poured into the aqueous phase containing a low concentration of poly (vinyl alcohol). The resulting o/w emulsion was then agitated for 90 min at room temperature. The microspheres were filtered, washed with water and dried. The encapsulation efficiency into ethylcellulose was highest for indomethacin followed by ibuprofen and ketoprofen. This order was inversely correlated to the aqueous solubility. All drugs were encapsulated because of their low water solubility and the efficiency of encapsulation was improved by increasing the drug loading and the polymer:organic solvent ratio. The drug release in aqueous solution at pH 7.4, 37°C under sink conditions was governed by microsphere size, drug loading and polymer composition. The release of indomethacin from ethylcellulose microcapsules was too slow and could be increased by using more permeable polymers or polymers blends. The rates of polymer and drug precipitation during microsphere formation depended upon the organic solvent selected. When chloroform was used as the solvent, the drug precipitated before the polymer and indomethacin crystals were observed on the microsphere surface. With methylene chloride, the polymer precipitated before the drug and no drug crystals were seen.

REFERENCES

Alex, R. and Bodmeier, R. (1990). *J. Microencapsulation* **7** (3), 347–355.

Anonymous (1988). *Res. Discl.* **294**, 788–789. Through *Chem. Abst.* **110**, 29013.

Arneodo, C. and Benoit, J. P. (1987). *Polym. Mater. Sci. Eng.* **57**, 255–259. Through *Chem. Abstr.* **107**, 178534.

Arneodo, C., Benoit, J. P. and Thies, C. (1986). *S.T.P. Pharma* **2** (15), 303–306, Through *Chem. Abstr.* **105**, 139395.

Arneodo, C., Baszkin, A., Benoit, J. P. and Thies, C. (1988a). *A.C.S. Symp. Ser.* **370** (Flavor Encapsulation), 132–147, Through *Chem. Abstr.* **109**, 229000.

Arneodo, C., Baszkin, A., Benoit, J. P., Fellous, R. and Thies, C. (1988b). *Colloids Surf.* **34** (2), 159–169. Through *Chem. Abstr.* **110**, 120978.

Arshady, R. (1990a). *Polymer Eng. Sci.* **30**, 905–914.

Arshady, R. (1990b). *Polymer Eng. Sci.* **30**, 915–924.

Badawi, A. A. and El Sayed, A. A. (1980). *J. Pharm. Sci.* **69**, 492–497.

Badawi, A. A., Shoukri, R. A. and El Zainy, A. A. (1991). *Bull. Fac. Pharm.* (Cairo Univ.) **29** (2), 61–68. Through *Chem. Abstr.* **116**, 46222.

Baichwal, M. R. and Abraham, I. A. (1980). *Indian J. Pharm. Sci.* **42** (2), 48–51. Through *Chem. Abstr.* **94**, 90226.

Bailey, F. E. Jr. and Callard, R. W. (1959). *J. Appl. Polym. Sci.* **1**, 373–374. Through Jizomoto (1985).

Bakan, J. A. (1980). In *Controlled Release Technologies: Methods, Theory and Applications*. Vol. II, Chapter 4, (A. F. Kydonieus, ed.) pp. 83–105. CRC Press, Boca Raton.

Bakan, J. A. (1986). In *The Theory and Practice of Industrial Pharmacy*, 3rd edn., (L. Lachman, H. A. Lieberman, and J. L. Kanig, eds), pp. 412–429. Lea & Febiger, Philadelphia.

Bardet, L., Maillols, H. and Maillols, J. (1969). *J. Chim. Phys. Physicochim. Biol.* **66** (7–8), 1399–1405, Through *Chem. Abstr.* **72**, 22058.

Barnett, M. K. and Zisman, W. A. (1959). *J. Phys. Chem.* **63** 1241–1246. Through Singh and Robinson (1988).

Bechard, S. and McMullen, J. N. (1986). *Int. J. Pharm.* **31**, 91–98.

Becher, P. (1965). *Emulsions: Theory and Practice*, Chapter 7, pp. 267–325. Reinhold, New York.

Benita, S. and Donbrow, M. (1980). *J. Coll. Interface. Sci.* **77**, 102–109.

Benita, S. and Donbrow, M. (1982). *J. Pharm. Sci.* **71**, 205–210.

Benita, S., Benoit, J. P., Puisieux, F. and Thies, C. (1984). *J. Pharm. Sci.* **73**, 1721–1724.

Benoit, J. P. and Puisieux, F. (1986). In *Polymeric Nanoparticles and Microspheres*, (P. Guiot and P. Couvreur, eds), Chapter 5, pp. 137–174. CRC Press, Boca Raton.

Benoit, J. P., Courteille, F. and Thies, C. (1986). *Int. J. Pharm.* **29**, 95–102.

Bergisadi, N. and Gurvardar, D. (1989). *Acta Pharm Turc.* **31** (4), 161–165. Through *Chem. Abstr.* **112**, 240377.

Berseneva, E. A., Pavlova, L. A. and Smagina, L. S. (1988). *Khim. Farm. Zh.* **22** (1), 78–80. Through *Chem. Abstr.* **108**, 156417.

Beyger, J. W. and Nairn, J. G. (1986). *J. Pharm. Sci.* **75**, 573–578.

Blow, A. M. J., Botham, G. M., Fisher, D., Goodall, A. H., Tilcock, C. P. S. and Lucy, J. A. (1978). *FEBS Lett.* **94**, 305–310, Through Jizomoto (1985).

Bodmeier, R. and Chen, H. (1989). *J. Controlled Release* **10**, 167–175.

Bodmeier, R. and McGinity, J. W. (1987a). *Pharm. Res.* **4** (6), 465–471.

Bodmeier, R. and McGinity, J. W. (1987b). *J. Microencapsulation* **4** (4), 289–297.

Bodmeier, R. and McGinity, J. W. (1987c). *J. Microencapsulation* **4** (4), 279–288.

Bodmeier, R. and McGinity, J. W. (1988). *Inter. J. Pharm.* **43**, 179–186.

Borue, V. Y. and Erukhimovich, I. Y. (1990). *Macromolecules* **23**, 3625–3632.

Boyd, G. E., Adamson, A. W. and Myers, L. S. (1947). *J. Am. Chem. Soc.* **69**, 2836–2848.

Brodin, A. F., Ding, S. and Frank, S. G. (1986). *PCT Int. Appl. WO.* 8603676 A1. Through *Chem. Abstr.* **105**, 139630.

Bungenberg de Jong, H. G. (1949a). In *Colloid Science.* Vol. II, Chapter VIII, (H. R. Kruyt, ed.) pp. 232–258. Elsevier, New York.

Bungenberg de Jong, H. G. (1949b). In *Colloid Science*, Vol II, Chapter X, (H. R. Kruyt, ed.), pp. 335–432. Elsevier, New York.

Burgess, D. J. (1990). *J. Coll. Interface Sci.* **140**, 227–238.

Burgess, D. J. and Carless, J. E. (1984). *J. Coll. Interface Sci.* **94**, 1–8.

Burgess, D. J. and Carless, J. E. (1985). *Inter. J. Pharm.* **27**, 61–70.

Burgess, D. J. and Carless, J. E. (1986a). *Inter. J. Pharm.* **32**, 207–212.

Burgess, D. J. and Carless, J. E. (1986b). In *Coulombic Interactions in Macromolecular Systems* (A. Eisenberg, E. F. Bailey, eds). ACS Symposium Series **302**. ACS, Washington, DC. Through Burgess (1990).

Burgess, D. J., Kwok, K. K. and Megremis, P. T. (1991a). *J. Pharm. Pharmacol.* **43**, 232–236.

Burgess, D. J., Marek, T. A., Kwok, K. K. and Singh, O. W. (1991b). *Polym. Prepr.* **32** (1), 600–601.

Calanchi, M. and Maccari, M. (1980). In *Controlled Release Technologies: Methods, Theory and Applications.* Vol. II, (A. F. Kyodenieus, ed.), pp. 107–116. CRC Press, Boca Raton.

Cameroni, R., Coppi, G., Forni, F., Iannuccelli, V. and Bernabei, M. T. (1985). *Boll. Chem. Farm.* **124** (9), 393–400, Through *Chem. Abstr.* **104**, 74950.

Cavalier, M., Benoit, J. P. and Thies, C. (1986). *J. Pharm. Pharmacol.* **38**, 249–253.

Chalabala, M. (1984). *Pharmazie* **39** (3), 643–644, Through *Chem. Abstr.* **102**, 84352.

Chattaraj, S. C., Das, S. K., Karthikeyan, M., Ghosal, S. K. and Gupta, B. K. (1991). *Drug Dev. Ind. Pharm.* **17** (4), 551–560.

Chemtob, C., Gruber, T. and Chaumeil, J. C. (1989). *Drug Dev. Ind. Pharm.* **15** (8), 1161–1174.

Chilvers, G. R. and Morris, V. J. (1987). *Carbohydr. Polym.* **7** (2), 111–120. Through *Chem. Abstr.* **107**, 79869.

Chilvers, G. R., Gunning, A. P. and Morris, V. J. (1988a). *Carbohydr. Polym.* **8** (1), 55–61, Through *Chem. Abstr.* **109**, 8313.

Chilvers, G. R., Gunning, A. P. and Morris, V. J. (1988b). *P.C.T. Int. Appl. WO* 8808747 A1, Through *Chem. Abstr.* **110**, 97490.

Chilvers, G. R., Gunning, A. P. and Morris, V. J. (1988c). GB 2204553 A1, Through *Chem. Abstr.* **110**, 137476.

Cho, C. S., Lee, C. M., Chae, K. H. and Kim, J. M. (1982). *Pollimo* **6** (1), 54–59, Through *Chem. Abstr.* **96**, 163504.

Chowdary, K. P. R. and Annapurna, A. (1989). *Indian J. Pharm. Sci.* **51** (2), 53–56.

Chowdary, K. P. R. and Rao, G. N. (1984). *Indian J. Pharm. Sci.* **46**, 213–215.

Chowdary, K. P. R. and Rao, G. N. (1985). *Indian Drugs* **23** (1), 58, Through *Chem. Abstr.* **104**, 39648.

Chukwu, A., Agarwal, S. P. and Adikwu, M. (1991). *S.T.P. Pharma. Sci.* **1** (2), 117–120, Through *Chem. Abstr.* **115**, 214728.

Chun, I. K. and Shin, D. W. (1988). *Yakhak Hoechi* **32** (1), 26–39, Through *Chem. Abstr.* **109**, 43366.

Considine, D. M. (ed.) (1983). *Van Nostrand's Scientific Encyclopedia*, 6th edn. Van Norstrand Reinhold, New York.

Das, S. K. and Palchowdhury, S. (1989). *J. Microencapsulation* **6** (1), 53–58.

David, J., Lefrancois, C. and Ridoux, C. (1988). EP 273823 A1, Through *Chem. Abstr.* **109**, 134997.

Deasy, P. B. (1984a). *Microencapsulation and Related Drug Processes*, Chapter 3, pp. 61–95. Dekker, New York.

Deasy, P. B. (1984b). *Microencapsulation and Related Drug Processes*, Chapter 4, pp. 97–117. Dekker, New York.

Deasy, P. B. (1984c). *Microencapsulation and Related Drug Processes*, Chapter 1, pp. 1–19. Dekker, New York.

Deasy, P. B. (1984d). *Microencapsulation and Related Drug Processes*, Chapter 2, pp. 21–60. Dekker, New York.

Deasy, P. B., Brophy, M. R., Ecanow, B. and Joy, M. M. (1980). *J. Pharm. Pharmacol.* **32**, 15–20.

Devay, A. and Racz, I. (1988). *J. Microencapsulation* **5** (1), 21–25.

Dharamadhikari, N. B., Joshi, S. B. and Manekar, N. C. (1991). *J. Microencapsulation* **8** (4), 479–482.

Dhruv, A. B., Needham, Jr. T. E. and Luzzi, L. A. (1975). *Can. J. Pharm. Sci.* **10** (2), 33–36.

Donbrow, M. (1992). In *Microcapsules and Nanoparticles in Medicine and Pharmacy*, (M. Donbrow ed.), Chapter 2, pp. 17–45. CRC Press, Boca Raton.

Donbrow, M. and Benita, S. (1977). *J. Pharm. Pharmacol.* **29**, suppl., 4p.

Donbrow, M., Benita, S. and Hoffman, A. (1984). *Appl. Biochem. Biotechnol.* **10**, 245–249.

Donbrow, M., Hoffman, A. and Benita, S. (1990). *J. Microencapsulation* **7** (1), 1–15.

D'Onofrio, G. P., Oppenheim, R. C. and Bateman, N. E. (1979). *Int. J. Pharm.* **2**, 91–99.

Dubernet, C., Benoit, J. P. and Puisieux, F. (1991). *Eur. J. Pharm. Biopharm.* **37** (1), 49–54.

Duquemin, S. J. and Nixon, J. R. (1985). *J. Pharm. Pharmacol* **37**, 698–702.

Duquemin, S. J. and Nixon, J. R. (1986). *J. Microencapsulation* **3** (2), 89–93.

Ecanow, B. (1988a). *Eur. Pat. App. EP* 256856 A2. Through *Chem. Abstr.* **109**, 61458.

Ecanow, B. (1988b). *Eur. Pat. App. EP* 274431 A2. Through *Chem. Abstr.* **110**, 237136.

Ecanow, B. (1991). *Eur. Pat. App. EP* 416575 A2. Through *Chem. Abstr.* **115**, 99288.

Ecanow, C. S. and Ecanow, B. (1983). *PCT. Int. Appl. WO* 8302228 A1. Through *Chem. Abstr.* **99**, 93720.

Ecanow, J., Ecanow, D. and Ecanow, B. (1990). *Biomat. Art. Cells, Art. Org.* **18** (2), 359–365.

Edomond, E. and Ogston, A. G. (1968). *Biochem. J.* **109**, 569–576, Through Jizomoto (1985).

El Egakey, M. A. and El Gindy, N. A. (1983). *Drug Dev. Ind. Pharm.* **9** (5), 895–908.

El Gindy, N. A. and El Egakey, M. A. (1981a). *Drug Dev. Ind. Pharm.* **7** (6), 739–753.

El Gindy, N. A. and El Egakey, M. A. (1981b). *Drug·Dev. Ind. Pharm.* **7** (5), 587–603.

Embleton, J. K. and Tighe, B. J. (1992). *J. Microencapsulation* **9** (1), 73–87.

Encina, G. G., Sanghvi, S. P. and Nairn, J. G. (1992). *Drug Dev. Ind. Pharm.* **18** (5), 561–579.

El Sayed, A. A., Badawi, A. A. and Fouli, A. M. (1982). *Pharm. Acta Helv.* **57** (2), 61–64.

El Shattawy, H., Kassem, A. and El Razzaz, M. (1991). *Drug Dev. Ind. Pharm.* **17** (18), 2529–2537.

El Shattawy, H. H., Kassem, A. A., Nouh, A. T. and El Razzaz, M. A. (1992). *Drug Dev. Ind. Pharm.* **18** (1), 55–64.

El Shibini, H. A., Abdel-Ghani, Y. M. and Nada, A. H. (1989). *Alexandria J. Pharm. Sci.* **3** (1), 56–62, Through *Chem. Abstr.* **111**, 140390.

Falyazi, B. G., Bobrova, L. E. and Izmailova, V. N. (1975). *Vestn. Mosk. Univ. Khim.* **16** (3), 327–330. Through *Chem. Abstr.* **84**, 6500.

Fanger, G. O., Miller, R. E. and McNiff, R. G. (1970). US 3531418, Through Deasy, P.B. (1984) *Microencapsulation and Related Drug Processes*, Chapter 4, pp. 97–117, Dekker, New York.

Fellows, C. T., Brown, G. and Haines, R. C. (1987). EP 247864 A2, Through *Chem. Abstr.* **108**, 156283.

Flory, P. J. (1953). *Principles of Polymer Chemistry*, Chapter 13, pp. 541–594. Cornel University, Ithaca.

Fong, J. W. (1988). In *Controlled Released Systems: Fabrication Technology*. Vol. 1, (D. S .T. Hsieh, ed.), Chapter 5, pp. 81–108. CRC Press, Boca Raton.

Fong, J. W., Nazareno, J. P., Pearson, J. E. and Maulding, H. V. (1986). *J. Controlled Release* **3**, 119-130.

Fong, J. W., Maulding, H. V., Visscher, G. E., Nazareno, J. P. and Pearson, J. E. (1987). *ACS Symp. Ser. 348, Controlled Release Technol. Pharm. App.* 214-230, Through *Chem. Abstr.* **108**, 26883.

Frank, S. G., Brodin, A. F., Chen, C. M. J. and Panthuvanich, S. (1985). *PCT Int. Appl. WO* 8500105 A1. Through *Chem. Abstr.* **102**, 209439.

Fukui, Y. (1978). JP 53125987, Through *Chem. Abstr.* **90**, 88433.

Gekko, K., Harada, H. and Noguchi, H. (1978). *Agric. Biol. Chem.* **42** (7), 1385-1388. Through *Chem. Abstr.* **89**, 159042.

Gladilin, K. L. (1973). *Probl Vozniknoveniya Sushchnosti Zhizni* (A. I. Oparin ed.), pp. 115-119. Moscow, USSR. Through *Chem. Abstr.* **82**, 166251.

Gladilin, K. L., Orlovskii, A. F., Evreinova, T. N. and Oparin, A. I. (1972). *Dokl. Akad. Nauk SSSR* **206** (4), 995-998 (Biochem). Through *Chem. Abstr.* **78**, 13504.

Gopal, E. S. R. (1968). In *Emulsion Science*, Chapter 1 (P. Sherman, ed.), pp. 2-75. Academic Press, London.

Gove, P. B. (ed.) (1963). *Webster's Third New International Dictionary*. Merriam, Springfield.

Gupta, R. G. and Agarwal, C. P. (1983). *East. Pharm.* **26**, 133-137.

Gutcho, M. H. (1979). *Microcapsules and Other Capsules*, pp. 3-22. Noyes Data, Park Ridge.

Harris, M. S. (1981). *J. Pharm. Sci.* **70**, 391-394.

He, F. and Hou, H. (1989). *Zhongguo Yiyao Gongye Zazhi* **20** (10), 462-464, Through *Chem. Abstr.* **112**, 125140.

Hecquet, B., Fournier, C., Depadt, G. and Cappelaere, P. (1984). *J. Pharm. Pharmacol.* **36**, 803-807.

Heya, T., Okada, H. and Ogawa, Y. (1988). EP 256726 A2, Through *Chem. Abstr.* **109**, 116065.

Hiestand, E. N. (1966). US 3242051, Through Fong, J. W. (1988) In *Controlled Release Systems: Fabrication Technology*. Vol. 1, Chapter 5, (D. S. T. Hsieh, ed.), pp. 81-108. CRC Press, Boca Raton.

Hiestand, E. N., Jensen, E. H. and Meister, P. D. (1970). US 3539465. Through *Chem. Abstr.* **74**, 23456.

Hoerger, G. (1975). US 3872024. Through *Chem. Abstr.* **83**, 116296.

Hui, H. W., Lee, V. H. L. and Robinson, J. R. (1987). In *Controlled Drug Delivery*, 2nd edn. (J. R. Robinson, V. H. L. Lee, eds), Chapter 9, pp. 373-432. Dekker, New York.

Huttenrauch, R., Fricke, S. and Zielke, P. (1986). *Pharmazie* **41** (9), 666-667.

Ikeya, H., Takei, K. and Matsumoto, S. (1974a). JP 49115083. Through *Chem. Abstr.* **82**, 141141.

Ikeya, H., Takei, K. and Matsumoto, S. (1974b). JP 49111875. Through *Chem. Abstr.* **82**, 141627.

Iler, R. K. (1973). US 3738957. Through *Chem. Abstr.* **79**, 67391.

Iler, R. K. (1974). *J. Coll. Interface Sci.* **51** (3), 388-393.

Irwin, W. J., Belaid, K. A. and Alpar, H. O. (1988). *Drug Dev. Ind. Pharm.* **14** (10), 1307-1325.

Ishizaka, T., Endo, K. and Koishi, M. (1981). *J. Pharm. Sci.* **70**, 358–363.

Ishizaka, T., Ariizumi, T., Nakamura, T. and Koishi, M. (1985). *J. Pharm. Sci.* **74** (3), 342–344.

Iso, M., Suzuki, I. and Omi, S. (1985a). *Zairyo Gijutsu* **3** (6), 287–293, Through *Chem. Abstr.* **105**, 154263.

Iso, M., Kando, T. and Omi, S. (1985b). *J. Microencapsulation* **2** (4), 275–287, Through *Chem. Abstr.* **104**, 193032.

Itoh, M. and Nakano, M. (1980). *Chem. Pharm. Bull.* **28**, 2816–2819.

Itoh, M., Nakano, M., Juni, K. and Sekikawa, H. (1980). *Chem. Pharm. Bull.* **28**, 1051–1055.

Iwata, M. and McGinity, J. W. (1992). *J. Microencapsulation* **9** (2), 201–214.

Izgu, E. and Doganay, T. (1976). *Ankara Univ. Eczacilik Fac. Mecm.* **6** (1), 54–87, through *Chem. Abstr.* **85**, 198127.

Jaffe, H. (1981). US 4272398. Through Fong, J. W. (1988) In *Controlled Release Systems: Fabrication Technology.* Vol. 1, Chapter 5 (D. S. T. Hsieh, ed.), pp. 81–108. CRC Press, Boca Raton.

Jalil, R. and Nixon, J. R. (1989). *J. Microencapsulation* **6** (4), 473–484.

Jalil, R. and Nixon, J. R. (1990a). *J. Microencapsulation* **7** (3), 297–325.

Jalil, R. and Nixon, J. R. (1990b). *J. Microencapsulation* **7** (2), 229–244.

Jalil, R. and Nixon, J. R. (1990c). *J. Microencapsulation* **7** (2), 245–254.

Jalil, R. and Nixon, J. R. (1990d). *J. Microencapsulation* **7** (3), 357–374.

Jalil, R. and Nixon, J. R. (1990e). *J. Microencapsulation* **7** (3), 375–383.

Jalsenjak, I., Nicolaidou, C. F. and Nixon, J. R. (1976). *J. Pharm. Pharmacol.* **28**, 912–914.

Jalsenjak, V., Peljnjak, S. and Kustrak, D. (1987). *Pharmazie* **42** (6), 419–420.

Jani, G. K., Chauhan, G. M., Gohel, M. and Patel, J. (1992). *Indian Drugs* **29** (10), 450–452, Through *Chem. Abstr.* **117**, 118418.

Jizomoto, H. (1984). *J. Pharm. Sci.* **73**, 879–882.

Jizomoto, H. (1985). *J. Pharm. Sci.* **74**, 469–472.

Kaeser-Liard, B., Kissel, T. and Sucker, H. (1984). *Acta Pharm. Tech.* **30** (4), 294–301.

Kagemoto, A., Murakami, S. and Fujishiro, R. (1967). *Makromol Chem.* **105**, 154–163, Through Jizomoto (1985).

Kaltsatos, V., Rollet, M. and Perez, J. (1989). *STP Pharma.* **5** (2), 96–102, Through *Chem. Abstr.* **110**, 237076.

Kasai, S. and Koishi, M. (1977). *Chem. Pharm. Bull.* **25**, 314–320.

Kassem, A. A. and El Sayed, A. A. (1973). *Bull. Fac. Pharm. Cairo Univ.* **12** (2), 77–90, Through *Chem. Abstr.* **85**, 37189.

Kassem, A. A. and El Sayed, A. A (1974). *Bull. Fac. Pharm. Cairo Univ.* **13** (1), 51–56, Through *Chem. Abstr.* **86**, 60499.

Kassem, A. A., Badawy, A. A., El Sayed, A. A. and El Mahrouk, G. M. (1975a). *Bull. Fac. Pharm. Cairo Univ.* **14** (2), 91–105, Through *Chem. Abstr.* **89**, 94950.

Kassem, A. A., Badawy, A. A. and El Sayed, A. A. (1975b). *Bull. Fac. Pharm. Cairo Univ.* **14** (2), 115–135, Through *Chem. Abstr.* **89**, 94951.

Kato, T. (1983). In *Controlled Drug Delivery.* Vol. II, Chapter 7 (S. D. Bruck, ed.) pp. 189–240. CRC Press, Boca Raton.

Kawashima, Y., Lin, S. Y., Kasai, A., Takenaka, H., Matsunami, K., Nochida, Y. and Hirose, H. (1984). *Drug Dev. Ind. Pharm.* **10** (3), 467–479.

Keipert, S. and Melegari, P. (1992). *P. Z. Wiss* **5** (2), 84–89, Through *Chem. Abstr.* **117**, 33566.

Khalil, S. A. H., Nixon, J. R. and Carless, J. E. (1968). *J. Pharm. Pharmacol.* **20**, 215–225.

Khanna, N. M., Gupta, S. K., Sarin, J. S., Singh, S. and Kanna, M. (1984). DE 3228533 A1, Through *Chem. Abstr.* **100**, 180113.

Kim, C. K., Hwang, S. W., Hwang, S. J. and Lah, W. L. (1989). *Yakche Hakhoechi* **19** (2), 99–107, Through *Chem. Abstr.* **111**, 201538.

Kitajima, M., Kondo, A., Asaka, S., Morishita, M. and Abe, J. (1971) *Ger. Offen.* 2105039, Through *Chem. Abstr.* **75**, 152682 (1971).

Knorr, D. and Daly, M. (1988). *Process Biochem.* **23** (2), 48–50, Through *Chem. Abstr.* **109**, 115935.

Koh, G. L. and Tucker, I. G. (1988a). *J. Pharm. Pharmacol.* **40**, 233–236.

Koh, G. L. and Tucker, I. G. (1988b). *J. Pharm. Pharmacol.* **40**, 309–312.

Koida, Y., Hirata, G. and Samejima, M. (1983). *Chem. Pharm. Bull.* **31**, 4476–4482.

Koida, Y., Kobayashi, M., Hirata, G. and Samejima, M. (1984). *Chem. Pharm. Bull.* **32**, 4971–4978.

Koida, Y., Kobayashi, M. and Samejima, M. (1986). *Chem. Pharm. Bull.* **34**, 3354–3361.

Kojima, T., Nakano, M., Juni, K., Inoue, S. and Yoshida, Y. (1984). *Chem. Pharm. Bull.* **32**, 2795–2802.

Kondo, A. (1979a). *Microcapsule Processing and Technology*, Chapters 8, 9, 10, (J. W. Van Valkenburg, ed.), pp. 70–120. Dekker, New York.

Kondo, A. (1979b). *Microcapsule Processing and Technology*, Chapter 2, (J. W. Van Valkenburg, ed.), pp. 11–17. Dekker, New York.

Kondo, A. (1979c). *Microcapsule Processing and Technology*, Chapters 3 and 4, (J. W. Van Valkenburg, ed.), pp. 18–34. Dekker, New York.

Kreuter, J. (1978). *Pharm. Acta Helv.* **53** (2), 33–39.

Kristl, A., Bogataj, M., Mrhar, A. and Kozjek, F. (1991). *Drug Dev. Ind. Pharm.* **17** (8), 1109–1130.

Ku, Y. S. and Chin, S. Y. (1989). *Yakhak Hoechi* **33** (3), 191–202, Through *Chem. Abstr.* **111**, 180609.

Ku, Y. S. and Kang, H. H. (1991). *Nonchong – Han'guk Saenghwal Kwahak Yonguwon* **48**, 109–128, Through *Chem. Abstr.* **116**, 241873.

Ku, Y. S. and Kim, H. Y. (1984). *Yakhak Hoechi* **28** (4), 223–229, Through *Chem. Abstr.* **102**, 12259.

Ku, Y. S. and Kim, S. O. (1987). *Yakhak Hoechi* **31** (3), 182–188, Through *Chem. Abstr.* **108**, 101280.

Ku, Y. S. and Kim, J. Y. (1989). *Yakhak Hoechi* **33** (5), 312–318, Through *Chem. Abstr.* **112**, 223212.

Kwong, A. K., Chou, S., Sun, A. M., Sefton, M. V. and Goosen, M. F. A. (1986). *J. Controlled Release* **4**, 47, Through Watts *et al.* (1990).

Kyogoku, N., Saeki, Y. and Fujikawa, S. (1988). JP 63258641 A2, Through *Chem. Abstr.* **111**, 213639.

Lenk, T. and Thies, C. (1986). *ACS Symp Ser. 302* (Coulombic Interact. Macromol. Syst.), 240–250, Through *Chem. Abstr.* **104**, 213806.

Lewis, D. H. and Tice, T. R. (1984). *Long Acting Contraceptive Delivery Systems*, (G. I. Zatuchni, ed.), pp. 77–95. Harper Row, Philadelphia. Through *Chem. Abstr.* **102**, 137733.

Lim, F. L. and Moss, R. D. (1981). *J. Pharm. Sci.* **70**, 351–354.

Lin, S. Y. (1985). *J. Microencapsulation* **2** (2), 91–101, Through *Chem. Abstr.* **104**, 136008.

Lin, S. Y. (1987). *J. Microencapsulation* **4** (3), 213–216.

Lin, S. Y. and Yang, J. C. (1986a). *J. Controlled Release* **3**, 221–228.

Lin, S. Y. and Yang, J. C. (1986b). *T'ai-wan Yao Hsueh Tsa Chih* **38** (3), 160–165. Through *Chem. Abstr.* **107**, 161496.

Lin, S. Y. and Yang, J. C. (1987). *J. Pharm. Sci.* **76**, 219–223.

Lin, S. Y., Yang, J. C. and Jiang, S. S. (1985). *T'ai-wan Yao Hsueh Tsa Chih* **37** (1), 1–10, Through *Chem. Abstr.* **103**, 92799.

Lu, B., Zhang, L. and Deng, Y. (1986). *Huaxi Yike Daxue Xuebao* **17** (4), 328–331, Through *Chem. Abstr.* **106**, 125771.

Luu, S. N., Carlier, P. F., Delort, P., Gazzola, J. and Lafont, D. (1973). *J. Pharm. Sci.* **62**, 452–455.

Luzzi, L. A. (1970). *J. Pharm. Sci.* **59**, 1367–1376.

Luzzi, L. A. (1976). In *Microencapsulation*. Chapter 17, (J. R. Nixon, ed.) pp. 193–206. Dekker, New York.

Luzzi, L. A. and Gerraughty, R. J. (1964). *J. Pharm. Sci.* **53**, 429–431.

Luzzi, L. A. and Gerraughty, R. J. (1967). *J. Pharm. Sci.* **56**, 634–638.

Luzzi, L. and Palmieri III, A. (1984). In *Biomedical Applications of Microencapsulation*, Chapter 1, (F. Lim, ed.), pp. 1–17. CRC Press, Boca Raton.

Madan, P. L. (1978). *Drug Dev. Ind. Pharm.* **4** (1), 95–116.

Madan, P. L. (1980). *Drug Dev. Ind. Pharm.* **6** (6), 629–644.

Madan, P. L., Luzzi, L. A. and Price, J. C. (1972). *J. Pharm. Sci.* **61**, 1586–1588.

Maharaj, I., Nairn, J. G. and Campbell, J. B. (1984). *J. Pharm. Sci.* **73**, 39–42.

Maierson, T. (1969). US 3436452, Through Deasy, P. B. (1984). *Microencapsulation and Related Drug Processes*, Chapter 2, pp. 21–60. Dekker, New York.

Makino, K., Arakawa, M. and Kondo, T. (1985). *Chem. Pharm. Bull.* **33**, 1195–1201.

Manekar, N. C., Puranik, P. K. and Joshi, S. B. (1991). *J. Microencapsulation* **8** (4), 521–523.

Manekar, N. C., Puranik, P. K. and Joshi, S. B. (1992). *J. Microencapsulation* **9** (1), 63–66.

Marrs, W. M. (1982). *Prog. Food Nutr. Sci.* **6**, 259–268, Through *Chem. Abstr.* **97**, 197108.

Marty, J. J., Oppenheim, R. C. and Speiser, P. (1978). *Pharm. Acta Helv.* **53** (1), 17–23.

Mathiowitz, E., Saltzman, W. M., Domb, A., Dor, P. and Langer, R. (1988). *J. App. Polymer Sci.* **35**, 755-774.

McMullen, J. N., Newton, D. W. and Becker, C. H. (1982). *J. Pharm. Sci.* **71**, 628-633.

McMullen, J. N., Newton, D. W. and Becker, C. H. (1984). *J. Pharm. Sci.* **73**, 1799-1803.

Merkle, H. P. and Speiser, P. (1973). *J. Pharm. Sci.* **62**, 1444-1448.

Meshali, M. M., El-Dien, E. Z., Omar, S. A. and Luzzi, L. A. (1989a). *J. Microencapsulation* **6** (3), 339-353.

Meshali, M. M., El-Dien, E. Z., Omar, S. A. and Luzzi, L. A. (1989b). *J. Microencapsulation* **6** (3), 355-360.

Mesiha, M. S. and El-Sourady, H. A. (1984). *Bull. Pharm. Sci. Assiut. Univ.* **7** (1), 109-121, Through *Chem. Abstr.* **104**, 193029.

Milovanovic, D. and Nairn, J. G. (1986). *Drug Dev. Ind. Pharm.* **12** (8 and 9), 1249-1258.

Moldenhauer, M. G. and Nairn, J. G. (1990). *J. Pharm. Sci.* **79**, 659-666.

Moldenhauer, M. G. and Nairn, J. G. (1991). *J. Controlled Release* **17**, 49-60.

Moldenhauer, M. G. and Nairn, J. G. (1992). *J. Controlled Release* **22**, 205-218.

Morishita, M., Inaba, Y., Kobari, S. and Fukushima, M. (1973). DE 2237206, Through *Chem. Abstr.* **78**, 137561.

Morishita, M., Inaba, Y., Fukushima, M., Hattori, Y., Kobari, S. and Matsuda, T. (1976). US 3960757, Through Fong, J. W. (1988). In *Controlled Release Systems: Fabrication Technology*. Vol. 1, Chapter 5, (D. S. T. Hsieh, ed.), pp. 81-108. CRC Press, Boca Raton.

Morris, N. J. and Warburton, B. (1980). *J. Pharm. Pharmacol.* **32**, Suppl., 24p.

Morris, N. J. and Warburton, B. (1982). *J. Pharm. Pharmacol.* **34**, 475-479.

Morse, L. D. Boroshok, M. J. and Grabner, R. W. (1978). US 4107072, Through Fong, J. W. (1988). In *Controlled Release Systems: Fabrication Technology*. Vol. 1, Chapter 5, (D. S. T. Hsieh, ed.), pp. 81-108. CRC Press, Boca Raton.

Mortada, S. A. M. (1981). *Pharmazie* **36** (6), 420-423.

Mortada, S. A. M. (1989). *Alexandria J. Pharm. Sci.* **3** (1), 1-5, Through *Chem Abstr.* **111**, 140389.

Mortada, S. A. M., Motawi, A. M., El Egaky, A. M. and El Khodery, K. A. (1987a). *J. Microencapsulation* **4** (1), 11-21.

Mortada, S. A. M., El Egaky, A. M., Motawi, A. M. and El Khodery, K. A. (1987b). *J Microencapsulation* **4** (1), 23-37.

Mortada, S. A. M., El Egaky, M. A., Motawi, A. M. and El Khodery, K. A. (1988). *J. Microencapsulation* **5** (4), 311-317.

Motycka, S. and Nairn, J. G. (1979). *J. Pharm. Sci.* **68**, 211-215.

Motycka, S., Newth, C. J. L. and Nairn, J. G. (1985). *J. Pharm. Sci.* **74**, 643-646.

Nagura, M., Konishi, J., Qing, M. Y. and Ishikawa, H. (1988). *Kobunshi Ronbunshu* **45** (7), 581-586, Through *Chem. Abstr.* **109**, 166043.

Nakajima, A. and Sato, H. (1972). *Biopolymers* **10**, 1345-1355.

Nakajima, T., Takashima, Y., Iida, K. I., Mitsuta, H. and Koishi, M. (1987). *Chem. Pharm. Bull.* **35** (3), 1201-1206.

Nasa, S. L. and Yadav, S. (1989). *East. Pharm.* **32** (381), 133-134.

Nath, B. S. (1973). *Indian J. Pharm.* **35** (1), 26-27.

Nath, B. S. and Borkar, S. D. (1979). *East. Pharm.* **22** (263), 125-128.

Nath, B. S. and Shirwaiker, A. A. (1977). *Indian J. Pharm.* **39** (1), 10-12.

Nikolaev, A. S. (1990). *Farmatsiya (Moscow)* **39** (5), 31-35, Through *Chem. Abstr.* **114**, 12134.

Nikolayev, A. S. and Rao, P. G. (1984). *Indian J. Pharm. Sci.* **46** (Mar-Apr), 84-88.

Ninomiya, Y. (1986). JP 61004527 A2, Through *Chem. Abstr.* **105**, 30056.

Nixon, J. R. (1985). *Endeavour* NS9 (3), 123-128.

Nixon, J. R. and Harris, M. S. (1986). In *Development of Drugs and Modern Medicines*, Chapter 41, (J. W. Gorrod, G. G. Gibson and M. Mitchard, eds.), pp. 425-434. Ellis Horwood, Chichester.

Nixon, J. R. and Matthews, B. R. (1976). In *Microencapsulation*, Chapter 15, (J. R. Nixon, ed.), pp. 173-183. Dekker, New York.

Nixon, J. R. and Wong, K. T. (1989). *Int. J. Pharm.* **50**, 205-212.

Nixon, J. R., Khalil, S. A. H. and Carless, J. E. (1966). *J. Pharm. Pharmacol.* **18**, 409-416.

Nixon, J. R., Khalil, S. A. H. and Carless, J. E. (1968). *J. Pharm. Pharmacol.* **20**, 528-538.

Noda, A., Yamaguchi, M., Aizawa, M. and Kumano, Y. (1989). EP 316054 A1, Through *Chem. Abstr.* **113**, 11921.

Noda, A., Aizawa, M., Kumano, Y. and Yamaguchi, M. (1992). *J. SCCJ* **25** (4), 223-231, Through *Chem. Abstr.* **117**, 33398.

Noro, S.I., Ishii, F. and Saegusa, K. (1985). *Chem. Pharm. Bull.* **33**, 4649-4656.

North, B. (1989). EP 339866 A2, Through *Chem. Abstr.* **112**, 39109.

Nozawa, Y. and Higashide, F. (1978). In *Polymeric Delivery Systems* (R. J. Kostelnik, ed.), Gordon & Breach Science, New York, Through Fong, J. W. (1988). In *Controlled Release Systems: Fabrication Technology*. Vol. 1, Chapter 5, (D. S. T. Hsieh, ed.), pp. 81-108. CRC Press, Boca Raton.

Ohdaira, H. and Ikeya, H. (1973). JP 48103468, Through *Chem. Abstr.* **80**, 122439.

Ohno, H. and Higano, M. (1992). *Nippon Kagaku Kaishi* **1**, 124-126, Through *Chem. Abstr.* **116**, 86175.

Oita, N., Enescu, L., Stanciu, C., Grigorescu, S. and Rosca, D. (1982). RO 80056 B, Through *Chem. Abstr.* **100**, 56866.

Okada, J., Kusai, A. and Ueda, S. (1985a). *J. Microencapsulation* **2** (3), 163-173.

Okada, J., Kusai, A. and Ueda, S. (1985b). *J. Microencapsulation* **2** (3), 175-182.

Okihana, H. and Nakajima, A. (1976). *Bull. Inst. Chem. Res. Kyoto Univ.* **54** (2), 63-71, Through *Chem. Abstr.* **85**, 143766.

Okor, R. S. (1988). *Int. J. Pharm.* **47**, 263-264.

Okor, R. S. (1989). *J. Macromol. Sci. Phys.* **B28** (3-4), 364-374, Through *Chem. Abstr.* **111**, 154839.

Okor, R. S. (1990). *J. Controlled Release* **12** (3), 195-200, Through *Chem. Abstr.* **113**, 120698.

Okor, R. S. (1991). *J. Appl. Polym. Sci.* **43** (7), 1391-1392, Through *Chem. Abstr.* **115**, 184382.

Okor, R. S., Otimenyin, S. and Ijeh, I. (1991). *J. Controlled Release* **16**, 349-354.

Omi, S., Umeki, N., Mohri, H. and Iso, M. (1991). *J. Microencapsulation* **8** (4), 465-478.

Oner, L., Yalabik-Kas, S., Cave, G. and Hincal, A. A. (1984). *Labo Pharma-Probl. Tech.* **346**, 690-693, Through *Chem. Abstr.* **102**, 172505.

Oner, L., Kas, H. S. and Hincal, A. A. (1988). *J. Microencapsulation* **5** (3), 219-223.

Onions, C. T. (ed.) (1933). *The Shorter Oxford English Dictionary*. OUP, London.

Oowaki, T., Tanaka, M., Uesugi, K., Kasai, M. and Kashino, M. (1988). JP 63287544, Through *Chem. Abstr.* **110**, 59214.

Oppenheim, R. C. (1986). *Polymeric Nanoparticles and Microspheres*, Chapter 1, (P. Guiot and P. Couvreur, eds), pp. 1-25. CRC Press, Boca Raton.

Overbeek, J. Th. G. and Voorn, M. J. (1957). *J. Cell Comp. Physiol.* **49** (Suppl. 1), 7, Through Burgess (1990).

Ozer, A. Y. and Hincal, A. A. (1990). *J. Microencapsulation* **7** (3), 327-339.

Pal, P. R. and Pal, T. K. (1986). *Indian Drugs* **24** (2), 101-104, Through *Chem. Abstr.* **106**, 219508.

Pal, P. R. and Pal, T. K. (1987). *Indian Drugs* **24** (9), 430-437, Through *Chem. Abstr.* **107**, 205034.

Pal, P. R. and Pal, T. K. (1988). *Acta Pharm. Technol.* **34** (4), 204-207.

Palmieri III, A. (1977). *Drug Dev. Ind. Pharm.* **3** (4), 309-314.

Paradissis, G. N. and Parrott, E. L. (1968). *J. Clin. Pharm.* **8**, 54-59.

Pepeljnjak, S., Penovski, D. and Jalsenjak, V. (1988). *Pharmazie* **43** (10), 728.

Peters, H. J. W., Van Bommel, E. M. G. and Fokkens, J. G. (1992). *Drug Dev. Ind. Pharm.* **18** (1), 123-134, Through *Chem. Abstr.* **116**, 113469.

Phares, R. E. and Sperandio, G. J. (1964). *J. Pharm. Sci.* **53**, 515-521.

Pongpaibul, Y., Price, J. C. and Whitworth, C. W. (1984). *Drug Dev. Ind. Pharm.* **10** (10), 1597-1616.

Puglisi, G., Giammona, G., Santagati, N. A., Carlisi, B., Villari, A. and Sampinato, S. (1992). *Drug Dev. Ind. Pharm.* **18** (9), 939-959.

Rak, J., Chalabala, M. and Vitkova, M. (1984). *Farm. Obz.* **53** (5), 205-217. Through *Chem. Abstr.* **101**, 97590.

Reyes, Z. (1965). US 3173878, Through Fong, J. W. (1988). In *Controlled Release Systems: Fabrication Technology*. Vol. 1, Chapter 5, pp. 81-108. CRC Press, Boca Raton.

Robinson, D. H. (1989). *Drug Dev. Ind. Pharm.* **15** (14-16), 2597-2620.

Rowe, J. S. and Carless, J. E. (1981). *PCT Int. Appl. WO* 8102976 A1, Through *Chem. Abstr.* **96**, 91669.

Rozenblat, J., Shlomo, M. and Garti, N. (1989). *J. Microencapsulation* **6** (4), 515-526.

Ruiz, J. M., Tissier, B. and Benoit, J. P. (1989). *Inter. J. Pharm.* **49**, 69-77.

Ruiz, J. M., Busnel, J. P. and Benoit, J. P. (1990). *Pharm. Res.* **7** (9), 928-934.

Safwat, S. M. and El Shanawany, S. (1989). *J. Controlled Release* **9**, 65-73.

Salib, N. N., El-Menshawy, M. A. and Ismail, A. A. (1978). *Pharm. Ind.* **40** (11a), 1230-1234.

Salib, N. N., El-Gholmy, Z. A. and Hagar, H. H. S. (1989). *Alexandra J. Pharm. Sci.* 3 (2), 137–141, Through *Chem. Abstr.* 112, 84070.

Samejima, M. and Hirata, G. (1979). DE 2834373, Through *Chem. Abstr.* 90, 192552.

Samejima, M., Hirata, G. and Koida, Y. (1982). *Chem. Pharm. Bull.* 30, 2894–2899.

Samejima, M., Hirata, G. and Ishibashi, T. (1984). EP 99109 A1, Through *Chem. Abstr.* 100, 126912.

Sampath, S. S., Garvin, K. and Robinson, D. H. (1992). *Inter. J. Pharm.* 78, 165–174.

Sanghvi, S. P. and Nairn, J. G. (1991). *J. Pharm. Sci.* 80, 394–398.

Sanghvi, S. P. and Nairn, J. G. (1992). *J. Microencapsulation* 9 (2), 215–227.

Sanghvi, S. P. and Nairn, J. G. (1993). *J. Microencapsulation* 10 (2), 181–194.

Santamaria, R., Hermann, G. and Loffredo, A. (1975). *Rend. Atti. Accad. Sci. Med. Chir.* 129, 121–130, Through *Chem. Abstr.* 85, 138915.

Sato, H. and Nakajima, A. (1974a). *Colloid Polym. Sci.* 252, 294–297.

Sato, H. and Nakajima, A. (1974b). *Colloid Polym. Sci.* 252, 944–948.

Sato, H. and Nakajima, A. (1974c). *Kyoto Daigaku Nippon Kagakuseni Kenkyusho Koenshu* 31, 37–46, Through *Chem. Abstr.* 82, 73598.

Sawicka, J. (1985). *Farm. Pol.* 41 (6), 311–315, Through *Chem. Abstr.* 104, 155892.

Sawicka, J. (1990). *Pharmazie* 45 (4), 264–265, Through *Chem. Abstr.* 113, 65191.

Schnoering, H. and Schoen, N. (1970). *German Pat.* 1900865, Through *Chem. Abstr.* 73, 99755.

Seiyaku, T. Co. (1980). JP 55105615, Through *Chem. Abstr.* 94, 20425.

Shchedrina, L. A., Berseneva, E. A., Gryadunova, G. P., Matveeva, S. A., Dontsova, G. I., Barsel, V. A., Chlenov, V. A. and Tentsova, A. I. (1983). *Farmatsiya (Moscow)* 32 (4), 24–29, Through *Chem. Abstr.* 99, 181398.

Shekerdzhiiski, R., Pham, C. T. and Titeva, S. (1988). *Farmatsiya (Sofia)* 38 (5), 16–25, Through *Chem. Abstr.* 110, 101696.

Sheorey, D. S. and Dorle, A. K. (1990). *J. Microencapsulation* 7 (2), 261–264.

Sheorey, D. S. and Dorle, A. K. (1991a). *J. Microencapsulation* 8 (1), 71–78.

Sheorey, D. S. and Dorle, A. K. (1991b). *J. Microencapsulation* 8 (1), 79–82.

Sheorey, D. S., Shastri, A. S. and Dorle, A. K. (1991). *Inter. J. Pharm.* 68, 19–23.

Shin, S. C. and Koh, I. B. (1989). *Yakche Hakhoechi* 19 (1), 29–37, Through *Chem. Abstr.* 111, 84013.

Shionogi and Co. (1982). JP 57171432 A2, Through *Chem. Abstr.* 98, 113712.

Shively, M. L. and McNickle, T. M. (1991). *Drug Dev. Ind. Pharm.* 17 (6), 843–864.

Shopova, S. and Tomova, V. (1982). *Farmatsiya (Sofia)* 32 (5), 32–37, Through *Chem. Abstr.* 98, 185506.

Siddiqui, O. and Taylor, H. (1983). *J. Pharm. Pharmacol.* 35, 70–73.

Singh, J. and Robinson, D. H. (1988). *J. Microencapsulation* 5 (2), 129–137.

Singh, J. and Robinson, D. H. (1990). *J. Microencapsulation* 7 (1), 67–76.

Singh, M., Bala, K. and Vasudevan, P. (1982). *Makromol Chem.* 183 (8), 1897–1903, Through *Chem. Abstr.* 97, 115272.

Singh, O. N. and Burgess, D. J. (1989). *J. Pharm. Pharmacol* **41**, 670–673.

Singla, A. K. and Nagrath, A. (1988). *Drug Dev. Ind. Pharm.* **14** (10), 1471–1479.

Sparks, R. E. (1984). In *Encyclopedia of Chemical Technology*. Vol. 15, 3rd edn, (D. Eckroth, ed.), pp. 470–493. Wiley, New York.

Spelser, P. (1976). In *Microencapsulation*, Chapter 1, (J. R. Nixon, ed.) pp. 1–11. Dekker, New York.

Spenlehauer, G., Veillard, M. and Benoit, J. P. (1986). *J. Pharm. Sci.* **75**, 750–755.

Spenlehauer, G., Vert, M., Benoit, J. P., Chabot, F. and Veillard, M. (1988). *J. Controlled Release* **7**, 217–229.

Spiegl, P. and Jasek, W. (1977). *Sci. Pharm.* **45**, 47–53, Through *Chem. Abstr.* **87**, 11530.

Spiegl, P. and Viernstein, H. (1988). *Sci. Pharm.* **56** (3), 175–179, Through *Chem. Abstr.* **110**, 44861.

Spittler, J., Mathis, C. and Stamm, A. (1977). *Expo-Congr Int. Technol. Pharm.* 1st, vol. 3, 119–125. Through *Chem. Abstr.* **90**, 43761.

Sprockel, O. L. and Prapaitrakul, W. (1990). *Inter. J. Pharm.* **58**, 123–127.

Sprockel, O. L. and Price, J. C. (1990). *Drug Dev. Ind. Pharm.* **16** (2), 361–376.

Stanaszek, W. F., Levinson, R. S. and Ecanow, B. (1974). *J. Pharm. Sci.* **63**, 1941–1943.

Sveinsson, S. J. and Kristmundsdottir, T. (1992). *Inter. J. Pharm.* **82**, 129–133.

Szretter, D. and Zakrzewski, Z. (1984a). *Farm. Pol.* **40** (5), 275–279. Through *Chem. Abstr.* **101**, 216340.

Szretter, D. and Zakrzewski, Z. (1984b). *Acta Pol. Pharm.* **41** (2), 241–247, Through *Chem. Abstr.* **120**, 225976.

Szretter, D. and Zakrzewski, Z. (1987a). *Acta Pol. Pharm.* **44** (3-4), 352–356. Through *Chem. Abstr.* **109**, 11679.

Szretter, D. and Zakrzewski, Z. (1987b). *Acta Pol. Pharm.* **44** (6), 555–559, Through *Chem. Abstr.* **109**, 236904.

Tainaka, K. (1979). *J. Phys. Soc. Japan* **46**, 1899, Through Burgess (1990).

Tainaka, K. (1980). *Biopolymers* **19**, 1289, Through Burgess (1990).

Takahashi, M., Kiyama, K., Arai, H. and Takizawa, M. (1989). JP 01270937 A2 Through *Chem. Abstr.* **112**, 140864.

Takeda, Y., Nambu, N. and Nagai, T. (1981). *Chem. Pharm. Bull.* **29**, 264–267.

Takenaka, H., Kawashima, Y. and Lin, S. Y. (1979). *Chem. Pharm. Bull.* **27**, 3054–3060.

Takenaka, H., Kawashima, Y. and Lin, S. Y. (1980a). *J. Pharm. Sci.* **69**, 513–516.

Takenaka, H., Kawashima, Y. and Lin, S. Y. (1980b). *Funtai Kogaku Kaishi* **17** (4), 179–183, Through *Chem. Abstr.* **93**, 101433.

Takenaka, H., Kawashima, Y. and Lin, S. Y. (1981). *J. Pharm. Sci.* **70**, 302–305.

Takruri, H., Ecanow, B. and Balagot, R. (1977). *J. Pharm. Sci.* **66**, 283–284.

Thies, C. (1973). *J. Colloid Interface Sci.* **44** (1), 133–141.

Thies, C. (1975). *Polymer Plast. Technol. Eng.* **5** (1), 1–22.

Thies, C. (1982). *CRC Crit. Rev. Biomed. Eng.* **8** (4), 335–383.

Tice, T. R. and Gilley, R. M. (1985). *J. Controlled Release* **2**, 343–352.

Torza, S. and Mason, S. G. (1970). *J. Colloid Interface Sci.* **33** (1), 67–83.

Ushiyama, H. (1979). JP 54132479, Through *Chem. Abstr.* **92**, 148045.

Van Oss, C. J. (1988-1989). *J. Dispersion Sci. Technol.* **9** (5&6), 561-573.

Veis, A. and Aranyi, C. J. (1960). *J. Phys. Chem.* **64**, 1203-1210. Through Burgess (1990).

Veis, A. (1970a). In *Biological Polyelectrolytes*, Chapter 4, (A. Veis, ed.), pp. 211-273. Dekker, New York.

Veis, A. (1970b). *Amer. Chem. Soc. Div. Org. Coatings Plast. Chem. Pap.* **30** (2), 219-223, Through *Chem. Abstr.* **76**, 158713.

Veis, A. (1975). *Amer. Chem. Soc. Div. Org. Coatings Plast. Chem. Pap.* **35** (1), 342-347.

Voorn, M. J. (1956). *Rec. Trav. Chim.* **75**, 317, 405, 427, 925. Through Nakajima and Sato (1972).

Wajnerman, E. S., Grinberg, W. J. and Tolstogusow, W. B. (1972). *Kolloid-Z. Poly.* **250** (10), 945-949, Through *Chem. Abstr.* **78**, 102313.

Wakiyama, N., Juni, K. and Nakano, M. (1981). *Chem. Pharm. Bull.* **29**, 3363-3368.

Wakiyama, N., Juni, K. and Nakano, M. (1982). *Chem. Pharm. Bull.* **30**, 3719-3727.

Watts, P. J., Davies, M. C. and Melia, C. D. (1990). *Crit. Rev. Therapeutic Drug Carrier Systems* **7** (3), 235-259.

Wen, S., Xiaonan, Y. and Stevenson, W. T. K. (1991a). *Biomaterials* **12**, 374-384.

Wen, S., Xiaonan, Y., Stevenson, W. T. K. and Alexander, H. (1991b). *Biomaterials* **12**, 479-488.

Wong, K. T. and Nixon, J. R. (1986). In *Development of Drugs and Modern Medicines*, Chapter 39, (J. W. Gorrod, G. G. Gibson, M. Mitchard, eds), pp. 409-420. Ellis Horwood, Chichester.

Yagi, H. (1986). *Baiotekunoroji Kenkyu Hokokusho*, 67-72. Through *Chem. Abstr.* **108**, 134007.

Yagi, H. (1987). *Baiotekunoroji Kenkyu Hokokusho*, 59-66. Through *Chem. Abstr.* **108**, 134008.

Yong, J. I. and Kim, O. N. (1988). *Yakche Hakhoechi* **18** (4), 187-195. Through *Chem. Abstr.* **110**, 218935.

Yoshida, N. H. (1972). US 3657144, Through Fong, J. W. (1988). In *Controlled Release Systems: Fabrication Technology*, Vol. 1, Chapter 5, (D. S. T. Hsieh, ed.), pp. 81-108. CRC Press, Boca Raton, 1988.

Yoshida, M., Takao, S. and Osawa, K. (1989). JP 01148338 A2. Through *Chem. Abstr.* **112**, 140735.

Zholbolsynova, A. S., Izmailova, V. N. and Piskareva, R. I. (1971). *Izv. Akad. Nauk Kaz. SSR. Ser. Khim.* **21** (3), 36-40. Through *Chem. Abstr.* **75**, 92974.

Zholbosynova, A. S., Kozlov, A. S. and Bekturova, E. A. (1988). *Izv. Akad. Nauk Kaz SSR. Ser. Khim.* (3), 57-60. Through *Chem. Abstr.* **109**, 151839.

PARTICLE AGGLOMERATION

H. G. Kristensen

Department of Pharmaceutics, The Royal Danish School of Pharmacy, Copenhagen, Denmark

INTRODUCTION

Agglomeration of particulate solids is a size enlargement method by which fine particles are formed into larger entities by mechanical agitation, usually in the presence of a liquid phase. Agglomeration is a generic technology with applications to a variety of processing industries. In the pharmaceutical industry the most important objectives and benefits of agglomeration processes are to convert mixtures of raw materials in powder form into formulations suitable for tabletting by creating non-segregating blends with improved flow and compaction properties, to decrease handling hazards from dust formation and to control the dissolution of drug substances from final dosage forms. Particle agglomeration is achieved by wet granulation with polymer binder solutions or by melt granulation with a softened or molten binder material which is solid at room temperature. Melt granulation is also known as thermoplastic granulation. Wet and melt granulation may also be used to prepare pelletized products, i.e. uniformly sized granules with preferably spherical shape, which are intermediate products in the manufacture of gastro-resistant and prolonged release formulations either in tablet or capsule form.

Research on wet and melt granulation has focused mainly on the performance of granulating equipment and effects of operating variables and material properties on granule growth and morphology. The results tend to be very apparatus- and product-sensitive. It is still difficult to make a good prediction of the liquid requirements of a particular process and it is difficult to predict the effects of even small changes in operating conditions

ADVANCES IN PHARMACEUTICAL SCIENCES
ISBN 0-12-032307-9

on the granule growth. Scale-up of granulation is often a haphazard undertaking.

Early efforts have identified the mechanisms of agglomeration and the governing forces in agglomerate formation and growth. In particular, the pioneering work in the 1960s by Rumpf and his collaborators on mobile liquid bondings has contributed greatly to the understanding of the fundamentals of particle agglomeration. There is, however, a lack of knowledge of the relationship between solid–liquid interactions on the micro-level and the agglomeration process on the macro-level.

The fundamentals of size enlargement methods, equipment and their application to different industries and various types of products are reviewed in monographs by Kapur (1978), Capes (1980), Sherrington and Oliver (1981) and Pietsch (1984). Pharmaceutical wet granulation is reviewed by Kristensen and Schaefer (1987, 1992). Industrial wet granulation, including production experience, is presented in a series of reviews edited by Lindberg (1988). Other important reviews are given by Fonner *et al.* (1982) on the characterization of granulations and interactions between granule and tablet characteristics, Ghebre-Selassie (1989) on pharmaceutical pelletization technology, and Kristensen (1988b) on binders for granulation and tabletting.

The aim of the present review is to discuss the fundamentals of particle agglomeration. Attention is given to agglomeration mechanisms and liquid requirements. It presents the effects of starting material properties and process conditions on granule growth in mixer-granulators, especially high shear mixers which have been the subject of the majority of the pharmaceutical granulation literature in recent years. For more specific information on production experience with different types of granulator the reader should refer to the reviews mentioned above.

MECHANICAL PROPERTIES OF AGGLOMERATES

Liquid states

Whenever a particulate solid is mixed with a liquid, which wets the solid surfaces, the three-phase system of solid, liquid and gas will tend to reduce its free energy by the formation of liquid bridges between the particles. The cohesive forces established by the liquid bridges may cause agglomeration and consolidation of the agglomerates insofar as they can resist the disruptive forces caused by the mechanical agitation of the moistened mass.

Figure 1 outlines the liquid states in agglomerates according to Newitt and Conway-Jones (1958). An assembly of uniformly sized, spherical par-

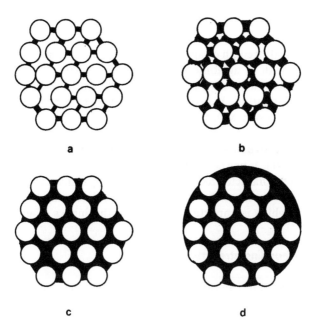

Fig. 1 Models for the state of liquid in a particle agglomerate. a, Pendular state; b, funicular state; c, capillary state; and d, droplet state.

ticles is wetted with increasing amounts of liquid. The relative amount of liquid is expressed by the liquid saturation, i.e. the volume of liquid relative to the volume of pores and voids between solid particles. At low liquid saturations (the pendular liquid state), discrete, lens-shaped liquid bridges are created between the particles. By increasing liquid saturation (the funicular liquid state), the liquid bridges coalesce and gradually fill the voids so that, at high liquid saturations (the capillary state), the particle agglomerate is held together by the capillary suction of the liquid. The limiting liquid saturation between the pendular and funicular liquid states is 25–35%. In the capillary liquid state the saturation exceeds about 80% (Capes, 1980).

The cohesive strength of the agglomerates described in Fig. 1 is due to the pressure deficiency caused by the surface tension of the liquid and the contact angle of the liquid to the solid surface. The strength of mobile liquid bondings was investigated and described by mathematical models for idealized systems by Rumpf and his co-workers (see below). A comprehensive presentation of models for liquid bondings can be found in most reviews on enlargement methods, for example Kapur (1978) and Pietsch (1984).

The mechanical strength of agglomerates has an important role in the formation and growth of agglomerates because the ability of an agglomerate to survive and grow must depend on its strength relative to disruptive forces resulting from agitation. It is difficult, however, to correlate agglomerate formation and growth with the bonding strength. The models for mobile liquid bondings mentioned below describe static systems and do not take into account the dynamic conditions present in agglomeration processes (Ennis *et al.*, 1991). When considering growth kinetics, the strength of agglomerates has to be characterized by more than one parameter. The tensile and shear strengths as well as the strain behaviour determine the agglomerate deformability which is important to the growth kinetics (Kapur, 1978).

Mobile liquid bonding strength

The tensile strength of agglomerates with localized bondings can be approximated by the following equation (Rumpf, 1962):

$$\sigma_t = \frac{9}{8} \cdot \frac{[1 - \epsilon]}{\epsilon} \cdot \frac{H}{d^2} \tag{1}$$

in which σ_t is the mean tensile strength per unit section area, ϵ is the void fraction of the agglomerate, d is the diameter of the particles, and H is the tensile strength of a single bond. The equation was derived by considering a particle assembly of monosized spheres.

An approximate solution of equation 1 for agglomerates in the pendular state was given by Pietsch (1969):

$$\sigma_t = \frac{9}{4} \cdot \frac{[1 - \epsilon]}{\epsilon} \cdot \frac{\gamma}{d} \tag{2}$$

The cohesive strength H of a pendular bonding depends on liquid surface tension γ, volume of liquid in the bridge, diameter d of the solid particles and the distance between the particles. Pietsch showed that H can be approximated by $2\gamma d$ when the particles are in close contact and when the contact angle of the liquid to the solid is zero. Equation 2 predicts that the tensile strength of an agglomerate in the pendular liquid state is constant and, hence, independent of the liquid saturation. The validity of the equation was demonstrated experimentally.

The tensile strength of agglomerates in the capillary state is controlled entirely by the pressure deficiency P in the liquid. P can be calculated from the Laplace equation for a circular capillary:

$$P = \frac{2\gamma}{r} \cos\theta \tag{3}$$

where r is the radius of the capillary and θ the contact angle.

Rumpf (1962) showed that the radius of the capillary may be related to the properties of the particle assembly by a hydraulic radius derived from the specific surface of the particles and the porosity of the assembly. Thus, the tensile strength due to the maximum pressure deficiency in an agglomerate of uniform spheres is:

$$\sigma_t = 6\frac{[1 - \epsilon]}{\epsilon} \cdot \frac{\gamma}{d} \cos\theta \tag{4}$$

Equations 2 and 4 show that the tensile strength of agglomerates in the pendular state is about one-third of the maximum strength in the capillary state. The tensile strength of agglomerates in intermediate states is usually approximated by the value of equation 4 multiplied by the liquid saturation S. Hence, the following equation describes the tensile strength of the agglomerate in the funicular and capillary liquid states:

$$\sigma_t = SC\frac{[1 - \epsilon]}{\epsilon} \cdot \frac{\gamma}{d} \cos\theta \tag{5}$$

C is a constant that takes the value 6 when the particles are uniform spheres. For irregular sand particles, values of C between 6.5 and 8 have been reported (Capes, 1980).

Figure 2 shows the theoretical tensile strength in the funicular and capillary state of an agglomerate consisting of $20\,\mu m$ spheres wetted with a binder liquid with a surface tension of $68\,\mathrm{mN\,m^{-1}}$, e.g. an aqueous povidone (PVP) solution. The graph demonstrates that, for example, an agglomerate with 20% porosity and a liquid content close to saturation has tensile strength of about $8\,\mathrm{N\,cm^{-2}}$ which is close to $1\,\mathrm{kg\,cm^{-2}}$.

The equations for the liquid bonding strength apply to idealized agglomerates consisting of uniform spheres. When the particle assembly is polydisperse, an approximate value of the tensile strength, resulting from mobile liquid bondings, can be obtained by substituting the volume–surface diameter d_{vs} of the particle system for d in the equations (Rumpf and Turba, 1964).

Equation 5 predicts that the tensile strength is proportional to porosity function $(1 - \epsilon)/\epsilon$. There is, however, experimental evidence for a larger variation of tensile strength with changes in porosity than that predicted by the equation. According to Cheng (1968), the major factors determining the tensile strength are the particle size distribution and the interparticle forces which are strongly dependent on the surface separation between

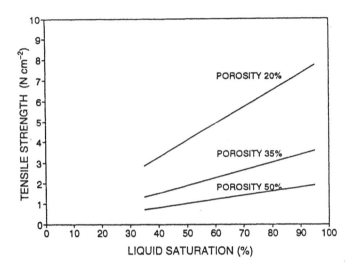

Fig. 2 Predicted tensile strength due to mobile liquid bonding of agglomerates with different porosities. Spheres of diameter 20 μm wetted with a liquid having surface tension 68 mN m^{-1} and zero contact angle to the solids.

solids. Tsubaki and Jimbo (1984) showed that the precompressive stress, applied to achieve a desired packing density of the particle assembly, influences the apparent tensile strength because of effects of interparticle forces in addition to effects of mobile liquid bondings. Although the dense particle packing achieved in agglomeration processes must be due to free movement of the particles into a tight packing, it is assumed that interparticle forces also affect the deformability of moist agglomerates composed of irregular particles with wide size distributions.

Compressive strength

Compressive strength is widely used in testing the tensile strength because of the simplicity of the test. In the case of a sphere with diameter D, the compressive strength σ_c is related to the fracture force P by the equation:

$$\sigma_c = \frac{4P}{\pi D^2} \tag{6}$$

The compressive strength of a sphere or a cylindrical sample loaded diametrically is greater than the corresponding tensile strength. In translating the compressive load into a tensile stress, the internal friction between moving particles must be overcome in addition to the tensile rupturing of the bonds in the fracture plane. Rumpf (1962) found that the ratio between tensile strength and compressive strength of moist limestone pellets took values between 0.5 and 0.8.

Figure 3 shows the compressive strength of cylindrical samples of commercial qualities of lactose and calcium hydrogen phosphate, both moistened with water. The particle size of the two powders is characterized by the geometric mean weight diameter, d_{gw}, corresponding to the median of the particle weight distribution. Assuming a log-normal weight distribution, the geometric standard deviation s_g expresses the ratio between diameters d_{84} and d_{50} corresponding to the 84% and 50% fraction of the weight distribution. The samples were prepared by very slow compaction in a tablet die of the moistened powders into cylinders of diameter 11.3 mm and length 4.6 mm. The investigated range of sample porosities is comparable with the intragranular porosities achieved by wet granulation of the powders. The strength of the compressed cylinders was measured by a diametrical compression test. Insofar as the samples were brittle, they fractured along the diameter parallel to the load. For lactose samples, the strength could not be measured at the higher liquid saturations because of a pronounced plastic deformation of the sample.

Figure 3 shows that the compressive strength of the moist samples is

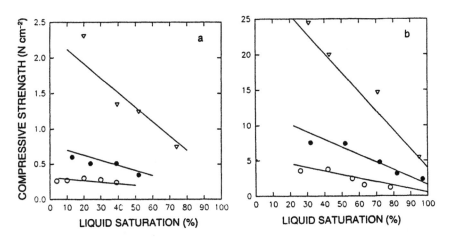

Fig. 3 Compressive strength of moist samples of lactose ($d_{gw} = 56\ \mu$m, $s_g = 1.8$) and calcium hydrogen phosphate ($d_{gw} = 14\ \mu$m, $s_g = 2.2$). a: Lactose; porosity 43% (○), 37% (●) and 30% (▽). b: Calcium hydrogen phosphate; porosity 50% (○), 37% (●) and 30% (▽). Reproduced with permission from Kristensen et al. (1985a), Powder Technol. **44**, 227–237. Elsevier Sequoia, NL.

dependent on porosity and liquid saturation. In the range of liquid saturations where the test samples are brittle, the compressive strength exceeds the tensile strength predicted by equation 5. Kristensen *et al.* (1985a) attributed this to effects of interparticle forces. The validity of the measurements was demonstrated by measuring the compressive strength of samples of moist glass spheres, which was found in reasonable agreement with equation 5. It appears, therefore, that the interparticle forces contributing to the strength of moist agglomerates, are significant in the case of powders with a wide size distribution; this is in agreement with the findings of Cheng (1968).

Experimental results on the tensile strength of moistened powders consisting of fine particles with a wide size distribution are presented by Eaves and Jones (1971, 1972a,b). Using the traction table method, they found that the tensile strength of sodium chloride and potassium chloride samples increased significantly when a small amount of water was added. Further addition produced decreasing tensile strength values. With calcium phosphate samples, the tensile strength remained constant until the mass was saturated with water. Their results relate to samples with porosities above 60%. The compressive strength of moist samples of calcium carbonate with porosities below 50% was measured by Takenaka *et al.* (1981) using a diametral compression test. They found that the compressive strength has a maximum at liquid saturations of 20–30%. Beyond this range the strength was reduced as the liquid saturation increased, which is similar to the effect shown in Fig. 3.

The results shown in Fig. 3 were analysed according to a model for the tensile strength of single powders and binary mixtures presented by Chan *et al.* (1983). For a single powder, the model takes the form:

$$\sigma_t = A \frac{[1 - \epsilon]}{\epsilon} \cdot \frac{\alpha}{t} \qquad (7)$$

where A is a constant, α is a material characteristic expressing the intrinsic interaction between like particles in a pair, and t is the distance of separation between particles. It was found that the distance parameter t was constant at a particular porosity, independent of the moistening liquid, and that α was constant at a particular liquid saturation. The solid lines in Fig. 3 represent the compressive strength predicted by equation 7 using the estimated values of α and t. Figure 4 shows the estimates of α against liquid saturation in experiments where the samples were moistened with water or an aqueous solution of a copolymer of polyvinylpyrrolidone (PVP) and polyvinylacetate (PVA); α, which has the dimensions of work, diminishes as the liquid saturation increases. By extrapolation, it appears that for lactose the effect of particle interactions disappears at complete saturation,

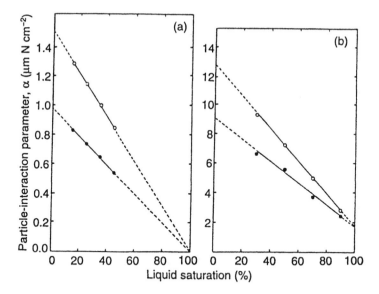

Fig. 4 Intrinsic interaction parameter α for lactose (a) and calcium hydrogen phosphate (b). ○, Samples wetted with water; ●, samples wetted with a 10% m/m solution of Kollidon VA64 in water. Reproduced with permission from Kristensen *et al.* (1985a), *Powder Technol.* **44**, 227–237. Elsevier Sequoia, NL.

while for calcium hydrogen phosphate some effect of the interparticle forces remains at complete saturation.

Figure 4 indicates that at medium liquid saturations, i.e. in the funicular liquid state, interparticle forces contribute significantly to the strength determined by compressive testing. As the liquid saturation is increased, the effects of the interparticle forces diminish so that the strength is expected to approach the strength resulting from mobile liquid bondings. It should be noted that the presence of a polymeric binder reduces the particle interactions.

Strain behaviour

Plastic deformation during collision between two agglomerates dissipates the kinetic energy of the system and thus improves the probability for a successful coalescence. Kapur (1978) claimed that an agglomerate with a small fracture strain is more prone to breakage than one with a large fracture strain, even when the tensile strength is the same in both cases. The stress–strain relationship is, therefore, one of the principal factors governing agglomerate growth by coalescence.

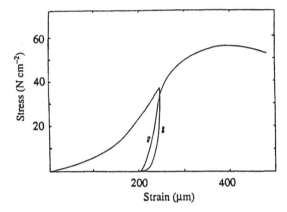

Fig. 5 Stress–strain relationship with cyclic loading and unloading for a moist sample of calcium hydrogen phosphate with porosity 45% and liquid saturation 57%. Reproduced with permission from Kristensen *et al.* (1985a), *Powder Technol.* **44**, 227–237. Elsevier Sequoia, NL.

Figure 5 shows the stress–strain relationship with cyclic loading and unloading for a cylindrical calcium hydrogen phosphate sample moistened with water and submitted to uniaxial stress. Although the sample behaved in a brittle way when exposed to the compressive stress, it is apparent that the strain is far from that of an ideal elastic–brittle body. The strain at stresses well below the fracture stress is partially plastic. Schubert (1975) has shown a similar strain behaviour in moist limestone samples exposed to tensile stresses.

Figure 6 shows the normalized strain of cylindrical samples of lactose and calcium hydrogen phosphate submitted to a uniaxial compressive test by applying a load to the end of the cylinder. The physical characteristics of the two powders were the same as those described in Fig. 3, but the moistening liquid used for the results presented in Fig. 6 was an aqueous PVP-PVA copolymer solution. The dotted lines in the two graphs indicate the condition when the sample becomes strained as a perfect plastic body, i.e. the strain is maintained at a constant stress. For lactose, the normalized strain increases with the liquid saturation, and perfect plastic strain is seen at saturations well below 100%, while for calcium hydrogen phosphate saturations close to 100% are required to achieve plasticity.

Figure 6 demonstrates that the moist samples change from being brittle into a state characterized by plastic deformation when the liquid saturation is increased to a certain range which depends on the packing density of the solid particles. Figure 7 compares the liquid saturations which produce

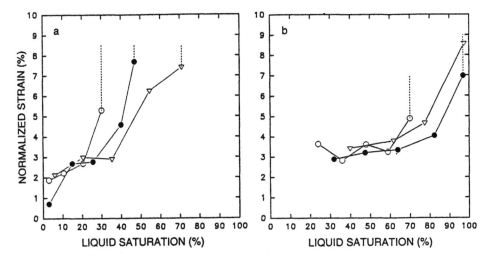

Fig. 6 Normalized strain $\Delta L/L$ of cylindrical samples of length L at maximum compressive stress. a: Lactose samples wetted with a 10% Kollidon VA64 solution in water, porosity 43% (○), 37% (●) and 30% (▽). b: Calcium hydrogen phosphate samples wetted with the same solution, porosity 50% (○), 43% (●) and 37% (▽). Reproduced with permission from Holm et al. (1985b), Powder Technol. **43**, 225–233. Elsevier Sequoia, NL.

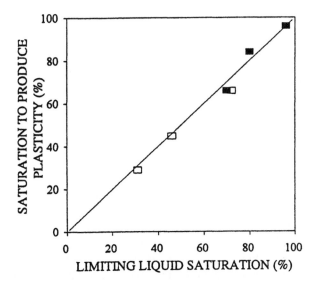

Fig. 7 Comparison of the liquid saturation required to achieve plastic deformation of moist samples and the limiting liquid saturations at which the compressive strength becomes controlled by mobile liquid bondings. □, Lactose; ■, calcium hydrogen phosphate.

plastic deformation (c.f. Fig. 6) with the liquid saturations at which the tensile strength caused by mobile liquid bondings (equation 5) equals the compressive strength shown in Fig. 3. Figure 7 demonstrates that the change from brittleness into plasticity occurs when the strength of the agglomerates becomes controlled entirely by mobile liquid bondings, i.e. when the interparticle forces become insignificant because of the 'lubricating' effect of the liquid.

The differences between the strength and strain behaviour of lactose and calcium hydrogen phosphate can partly be attributed to differences in the cohesional strength of the two systems, which in turn is dependent on the inherent physical character of the solid and, particularly, to the different particle size distributions. In general, the finer the particles, the greater become the tensile and compressive strengths. The liquid requirements to achieve plastic strain are supposed to be influenced by the size dispersion of the particles and their shape. However, according to experiences with melt granulation of lactose, the aqueous solubility of lactose may also influence the liquid requirements to achieve plasticity (see section on process and product variables).

AGGLOMERATE FORMATION AND GROWTH

Agglomeration mechanisms

Early work reviewed by Capes (1980) and Sherrington and Oliver (1981) has established that the following mechanisms for agglomerate formation and growth apply to many granulating systems:

1. Nucleation of primary particles by random coalescence
2. Coalescence between colliding agglomerates
3. Layering of primary particles or fines from degradation of established agglomerates
4. 'Ball growth'.

Nucleation of primary particles is a mechanism common to all agglomeration processes. Beyond or parallel to the nucleation stage, granule growth may proceed by coalescence between agglomerates when the starting material has a wide size distribution, or by the layering mechanism when the size distribution is narrow. Balling of agglomerates may be seen in mixer–granulators where larger agglomerates roll and slide over inclined surfaces. Kristensen (1988a) claimed that ball growth is essentially uncontrolled coalescence which proceeds in granulation regimes where shear effects are absent or low.

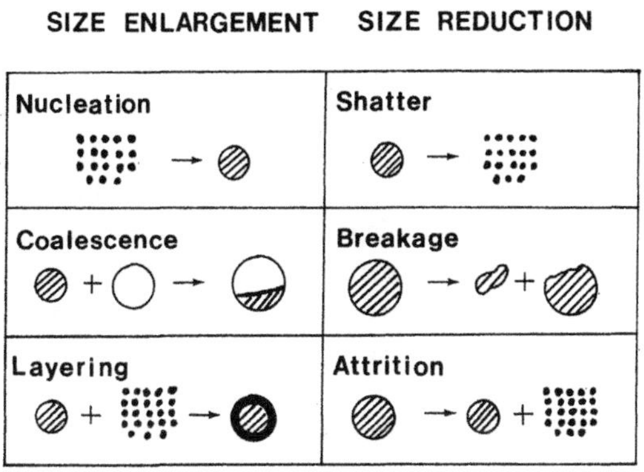

SIZE ENLARGEMENT   SIZE REDUCTION

Fig. 8 Formal presentation of mechanisms for size changes in an agglomeration process. Adapted from Sastry and Fuerstenau (1977).

Figure 8 shows the mechanisms of size enlargement and size reduction which are likely to control most agglomeration processes. Agglomerate growth proceeds in a balance between size enlargement and size reduction, which means a balance between binding and disruptive forces. What happens in a particular process depends on the very complex interactions between factors related to the granulator and its mode of action, the process conditions, and the properties of the powder to be agglomerated. Clearly, the strength of the moist agglomerates and their ability to resist the disruptive forces resulting from agitation of the mass are important to the overall growth process.

In wet granulation of pharmaceutical products, the starting materials are usually fine powders with wide particle size distributions which produce strong agglomerates that grow primarily by nucleation and coalescence. A uniform liquid distribution is a prerequisite for controlled growth. Good wetting properties of the binder liquid are, therefore, essential. Furthermore, it might be advantageous to add the binder solution slowly, by atomization, in order to avoid local over-wetting which gives rise to the presence of larger lumps in the final product, especially when the powders are soluble in the liquid.

Collision and coalescence

Figure 9 shows schematically a collision between two particles which may result in rebound or coalescence. In coalescence, the particles stick together because a bonding strength sufficient to resist the separating forces has been established. If the particles are smaller than a few micrometres, van der Waals forces of attraction suffice to overcome the competitive effects of gravity and kinetic energy of the particles. This means that aggregates are formed by agitation of the dry powder (Ho and Hersey, 1979). Normally, there must be free surface liquid present to agglomerate by liquid bonding. It is likely that the primary particles of the powder exhibit elastic deformation by impact so that the established pendular bondings must supply the bonding strength required to absorb the relative kinetic energy of the particles. In contrast, moist agglomerates may show plastic deformation by the impact through which kinetic energy is dissipated partly or fully by the deformation. This means a lower bonding strength is required to achieve coalescence of plastically deformable bodies compared with coalescence of elastic-brittle bodies of the same size.

The pendular bonding strength described by equation 2 shows that the strength is dependent on agglomerate porosity and surface tension. The equation applies to static conditions while the collision outlined in Fig. 9 is dynamic. Experience from production practice shows that wet granulation of a 'difficult' formulation may be facilitated by increasing the concentration of the binder liquid or by changing to a binder with a higher molecular weight, i.e. by applying a more viscous binder liquid.

It is well established that the viscosity influences the adhesion forces exerted by a liquid in dynamic conditions (Bowden and Tabor, 1964). This can be illustrated by reference to a model with two flat, parallel and circular plates separated by a film of liquid. Suppose the liquid is an 8% m/m solution of PVP K90 which has a surface tension 68 mN m^{-1} and viscosity of 100 mPa s (Ritala *et al.*, 1986). If the distance between the plates is 0.1 μm and the area of contact is 1 cm^2, the adhesive force between the two plates due to the surface tension is 272 N (or about 28 kg), c.f. equation 3.

The force F required to separate the plates within a certain time, t, is given by the following equation (Bowden and Tabor, 1964):

$$F = \frac{3\pi\eta R^2}{4t}\left[\frac{1}{h^2} - \frac{1}{k^2}\right]$$

(8)

where h is the initial distance between the plates and k is the final distance. In case of complete separation, i.e. $1/k^2 = 0$, equation 8 predicts that the force required to separate the plates ($R = 0.564$ cm) within 1 s is about 23 800 N. The apparent adhesive force is, thus, increased by a factor of 87 compared with the adhesive force in the static state. The internal strength of the liquid probably cannot resist the required tensile force so it will rupture at lower forces.

The example illustrates that the strength of liquid bondings under dynamic conditions may be significantly higher than the strength at static condi-

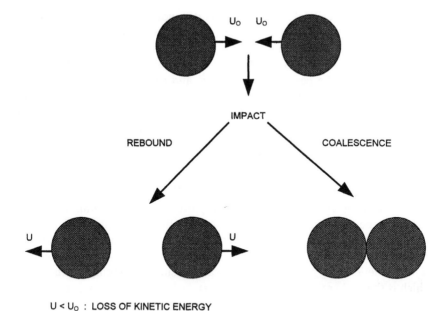

Fig. 9 Collision between two particles with relative velocity $2u_0$. The collision results in rebound with relative velocity $2u$ ($u < u_0$) or coalescence due to bonding forces created by free surface liquid.

tions. The factor by which the cohesive force is increased by the change depends on the viscosity of the liquid and the relative velocity of the solid particles.

The described effect of the liquid viscosity is likely to affect the growth rate in agglomeration processes. There is, however, only limited evidence for the effect. Ritala *et al.* (1986) have, in a comparison between different binders, shown that the binder concentration has a slight effect upon the granule growth in wet granulation in a high shear mixer, but the effect was much smaller than that of the liquid surface tension.

Ennis *et al.* (1991) have recently provided an interesting discussion of the coalescence between colliding particles. They showed that the cohesive force of the moving pendular liquid bridge comprises a capillary force component and a viscous force component, the latter being dominant. The prerequisite to a successful coalescence is that the dimensionless Stokes' number St_v, defined as:

$$St_v = \frac{2mu_o}{3\pi\eta r^2} \tag{9}$$

does not exceed a critical value, which is dependent on the strength of the pendular bond created by the collision. In equation 9, m is the mass of the particle, r is the particle diameter, u_o is the initial particle velocity and η is the viscosity of the liquid. St_v equals the ratio between the relative kinetic energy of the particles ($\frac{1}{2}m[2u_o]^2$) and the work done by the pendular bond at rebound ($F_{vis}2h$ where F_{vis} is the viscous bonding force and h is the thickness of the surface film on the particle). Intuitively, one can see that if the pendular bond can supply energy that exceeds the kinetic energy of the particles, the bond strength suffices to keep the particles together. Granulation regimes that apply to different granulation methods can be established on the basis of the Stokes' number.

The collision between particles outlined in Fig. 9 implies that the particle surfaces recover elastically, and may separate without residual deformation, as is the case when the particles are crystalline materials. With particle agglomerates, it is more likely that the collision is not truly elastic so that the impact results in a plastic deformation of the surfaces. The dissipation of the kinetic energy of the agglomerates results in heating of the mass which is the basic reason for the close correlation between power consumption and granule growth in high shear mixers (Holm *et al.*, 1985a).

Nucleation of particles

Agglomerate formation and growth by the nucleation mechanism proceed when there is sufficient free surface liquid to establish pendular bondings between the particles. It results in small, loose agglomerates which may grow further by coalescence with free particles. Insofar as the loose agglomerates survive the agitation they are likely to consolidate because of agitation and effects of the liquid bondings and, thereby, gain further strength.

Nucleation of particles is influenced by the cohesive forces expressed by equation 5 and probably also by the viscous forces of the moving pendular bridge. According to equation 5, reduction of the particle size gives rise to increased bonding strength. This is the reason why fine powders agglomerate more easily than coarse powders.

It is a general experience that for efficient agglomeration the binder liquid must have good wetting properties on the solid particles, but there are few data on the effect of the contact angle upon agglomerate growth. Ritala *et al.* (1988) found that granulation of fine particulate sulfur was difficult using a binder liquid with a contact angle of 56°. When the contact angle was reduced to 37°, by addition of a wetting agent, considerably less liquid was required but the granulation was still unsatisfactory. The effect of the contact angle is difficult to interpret, because the addition of a wetting agent changes both the surface tension and the contact angle. The literature on wet granulation gives the impression that a contact angle lower than about 25° in most experiments results in acceptable agglomeration.

The literature on binders for granule and tablet formulations shows that significant determinants for optimum granulation are the wetting of the solid by the binder and the binder adhesion and cohesion (Krycer *et al.*, 1983). The relative influence of these factors was assessed by Rowe (1988, 1989a–c, 1990) in studies on the thermodynamic energy of cohesion and adhesion. Although these studies relate to the properties of granules and tablets, they demonstrate that wetting and spreading of the binder influence the granule morphology. It is probable that the binder–substrate interaction influences the formation and growth of agglomerates. Parker *et al.* (1990, 1991) showed that binder–substrate interactions influence the rheological behaviour of microcrystalline cellulose wetted with different binder solutions. In addition to the effect of binder concentration and, hence, liquid viscosity, the spreading coefficient also affected the maximum torque recorded during wet massing. Different molecular weight grades of the polymers showed different torque readings at equivalent viscosities, indicating that the interactions between binder and substrate are dependent on the grade of the polymeric binder. In an earlier study, Jäger and Bauer (1984) demonstrated the benefit of using blends of low and high molecular weight grades of povidone as granule binder. Parker *et al.* (1990) found that increasing binder concentration and, hence, viscosity produced a greater maximum torque and also a reduced liquid requirement at the maximum. This agrees well with the 'lubricating' effect of adding a binder to the granulation, as shown in Fig. 4. The binder reduces the particle interactions and, therefore, improves the deformability of the moist agglomerates.

Figure 10 shows the final granule size obtained by wet granulation of lactose with various binder liquids in a fluidized bed granulator, i.e. a process characterized by low shear effects. As pointed out by Ennis *et al.* (1991), the resulting granule size is almost the same as the mean size of the droplets of the atomized binder liquid. As the binder droplet is deposited onto the bed it immediately absorbs surrounding particles. Because the particles have insufficient energy to rebound or break away, the drop structure is maintained in the formation of the agglomerate. Granule growth by fluidized bed granulation proceeds primarily by the nucleation mechanism insofar as there are primary particles present.

The effect of the droplet size shown in Fig. 10 can explain the many reports in the literature on effects of the binder liquid viscosity upon the final granule size in fluidized bed granulation (Kristensen and Schaefer, 1987). A high liquid viscosity gives rise to relatively large drops by atomization and, hence, a relatively large granule size.

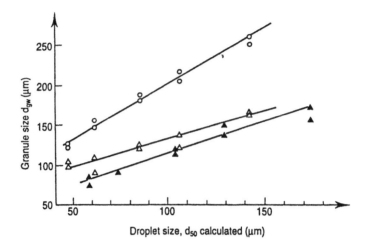

Fig. 10 Effect of mean droplet size on the final granule size in granulation of lactose with aqueous binder solutions in a fluidized bed granulator (Glatt WSG 15). Binder solutions: gelatine 4% (O); Povidone K25 10% (△); methylcellulose 2% (▲). Reproduced from Schaefer and Woerts (1978a) with permission from the authors.

In high shear mixers, growth by coalescence is significant in addition to the nucleation mechanism. The drop size of the atomized binder liquid does not affect the granule size (Holm *et al.*, 1983), as the particles absorbed by the drops break away because of the intensive agitation.

Coalescence of agglomerates

In the literature it is generally agreed that successful coalescence of agglo-merates occurs only when the agglomerates have excess surface liquid, mak-ing the surface plastically deformable. Surface plasticity is also required to round the resulting larger agglomerate. The excess surface liquid is sup-plied during the liquid addition phase of the process or, in the subse-quent wet massing phase, by liquid forced onto the agglomerate surface by consolidation.

According to the analysis by Ouchiyama and Tanaka (1982) of the coalescence mechanism, the probability for a successful coalescence bet-ween two colliding agglomerates is size dependent. Because of the mass and, hence, the kinetic energy of the colliding agglomerates, there must be an upper size limit beyond which coalescence is impossible. Ouchiyama and Tanaka derived the limiting agglomerate size δ:

$$\delta = A[K^{3/2}\sigma_c]^a \tag{10}$$

where A and a are constants, σ_c is the compressive strength and K is a parameter expressing the deformability of the agglomerate. Coalescence of particles larger than δ will not occur because the separating forces due to the kinetic energy exceed the binding forces. This corresponds to the situation described by Ennis *et al.* (1991) where the Stokes' number St_v in equation 9 exceeds the critical number because of increasing particle mass. Kristensen *et al.* (1985b) suggested that a high value of δ is associated with a high probability of achieving coalescence at random collision and, hence, a high growth rate.

The parameter K of equation 10 is related to compression force P acting between the two agglomerates and the area of contact A_s; $K = A_s/P$. For a small deformation of length Δl of a sphere, the area of contact is approximately $A_s = \pi D \Delta l / 2$, where D is the diameter of the spherical agglomerate. Hence, K can be expressed as:

$$K = \frac{\pi D \Delta l}{2P} \tag{11}$$

The compressive strength of the agglomerate equals the crushing force divided by the projected area of the agglomerate, i.e. $\sigma_c = 4P/(\pi D^2)$, c.f. equation 6. Inserting this equation and equation 11 into equation 10 and rearranging gives (Kristensen *et al.*, 1985b):

$$\delta^{2/a} = A_1 \frac{(\Delta l/D)^3}{\sigma_c} \tag{12}$$

where A_1 and a are constants and $\Delta l/D$ is the normalized strain of the agglomerate caused by the compression force P.

Equation 12 expresses the effect of agglomerate deformability upon the rate of growth by coalescence and presents the physical prerequisites for granule growth. The numerator expresses the strain produced by impact. As discussed earlier in the section on strain behaviour, the strain depends primarily on the packing density of the particles and the liquid saturation. Significant strain arises when the liquid saturation is increased to the limit where the cohesive strength of the agglomerate is governed by the strength of mobile liquid bondings as expressed by equation 5. The denominator of equation 12 can, therefore, be substituted by the tensile strength σ_t (equation 5).

Equation 12 predicts that a reduction of the particle size reduces the rate of growth by coalescence because of the effect of particle size on σ_c and σ_t. It can be compensated for by increasing the liquid saturation so that the strain behaviour is improved. The effect of improved strain will, according to equation 12, overrule the counteracting effect of the tensile strength because the normalized strain $\Delta l/D$ in the equation is raised to its third power. This can be achieved by increasing the amount of granulating liquid and/or increasing the consolidation of the agglomerates. This agrees well with the general experience that the finer the particles of the starting materials, the greater the amount of binder liquid required for efficient granulation.

The main implication of equation 12 for agglomeration processes is that the rate of growth by the coalescence mechanism is controlled primarily by the liquid saturation S given by the expression:

$$S = H\rho \frac{1 - \epsilon}{\epsilon} \qquad (13)$$

where H is the moisture content (humidity on dry basis) and ρ is the density of the solid. The equation assumes that the particles are insoluble in the liquid, and that the liquid has unit density.

Equation 13 shows that the liquid saturation is controlled by the amount of liquid phase present in the moistened powder and the porosity of the agglomerates. Because of the effect of the liquid saturation on agglomerate growth, consolidation of agglomerates during the process must have a pro-

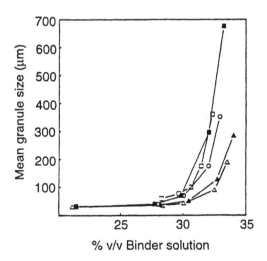

Fig. 11 Effect of binder solution upon granule growth in the liquid addition phase of granulation of calcium hydrogen phosphate in a Fielder PMAT 25 high shear mixer. Aqueous binder solutions: Kollidon 90, 3% (■); Kollidon VA64, 10% (○); Kollidon 25, 3% (□); Methocel E5, 3% (△); Methocel E15, 2% (▲). Reproduced with permission from Ritala *et al.* (1988), *Drug Dev. Ind. Pharm.* **14**, 1041–1060, Marcel Dekker Inc. USA.

nounced effect upon the growth rate. This is important, especially for wet granulation in high shear mixers where the intensive agitation may give rise to a pronounced densification of the agglomerates. Jaegerskou *et al.* (1984) found that the intragranular porosity of calcium hydrogen phosphate ($d_{gw} = 14\,\mu m$) was reduced steadily in the wet massing phase reaching a final value of about 20%. In contrast, granulation of the less cohesive lactose ($d_{gw} = 52\,\mu m$) showed that the final granule porosity was achieved early in the wet massing phase.

Effects of liquid saturation on growth

Figure 11 shows the effect of the amount of aqueous binder solution on the mean granule size achieved in the liquid addition phase when granulating calcium hydrogen phosphate in a high shear mixer. The data demonstrate that the type of binder has an effect upon the granule growth. In particular, povidone solutions give rise to greater granule sizes than solutions of hydroxypropylmethylcellulose (HPMC). Granulation with the povidone solutions produced denser granules than did the other binder solutions. Ritala *et al.* (1988) attributed the effect to a high surface tension, c.f. Table 1, which according to equation 5 produces a high liquid bonding strength. As a consequence, consolidation of the moist agglomerates becomes more pronounced.

Figure 12 shows the correlation between liquid saturation and mean granule size d_{gw} obtained by the experiments shown in Fig. 11 and additional experiments with concentrated solutions of the binders having viscosities up to about 100 mPa s. The viscosity had only a slight effect upon the granule growth.

The close correlation shown in Fig. 12 confirms equation 12 in predicting that the liquid saturation is a primary factor controlling agglomerate growth. The range of liquid saturations exceeds 100% due to a slight bias

Table 1 Viscosity and surface tension of the binder solutions shown in Fig. 11

Binder	Viscosity (mPa s, 30°C)	Surface tension (mN m^{-1}, 25°C)
Kollidon 90, 3%	9	68
Kollidon 25, 3%	1	68
Kollidon VA64, 10%	4	50
Methocel E5, 3%	6	48
Methocel E15, 2%	11	50

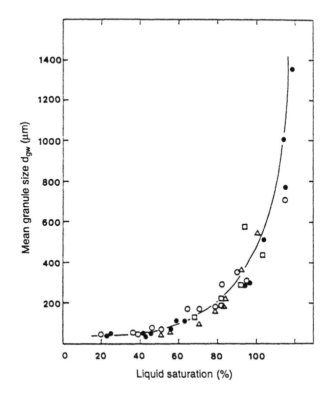

Fig. 12 The correlation between liquid saturation and the geometric mean weight diameter obtained in the same series of experiments as shown in Fig 11. Kollidon 90, 3%, 5%, 8%, ●; Kollidon VA64, 10%, 20%, 30%, ○; Kollidon 25, 3%, 20%, ▽; Methocel E5, 3%, 6%, 8%, △; Methocel E15, 2%, 3.5%, 4.5%, □. Reproduced with permission from Ritala *et al.* (1988), *Drug Dev. Ind. Pharm.* **14**, 1041–1060. Marcel Dekker Inc., USA.

in the determination of the intragranular porosity by a mercury immersion method, Jaegerskou *et al.* (1984). During measurement of the granule density, mercury may penetrate the surface of the granules giving rise to a density value that is too high. An error of, say, 2% in the intragranular porosity calculated from the density determination produces a systematic error of about 10% in the liquid saturation value.

Figure 12 demonstrates that significant agglomerate growth of calcium hydrogen phosphate requires liquid saturations close to 100%. This agrees well with the strain behaviour of the material described in Fig. 6. Complete saturation is required to achieve plasticity when the porosity is in the range of 20–30%, as is the case in the described granulation experiments.

Figure 6 also shows that moistened lactose becomes highly deformable at liquid saturations well below saturation. Kristensen *et al.* (1984) have, accordingly, found that wet granulation of lactose proceeds at liquid saturations below about 60%. Figure 13 shows the effect of the liquid saturation upon the mean granule size obtained by granulating lactose ($d_{gw} = 56\,\mu m$) in a high shear mixer. Examination of the moist agglomerates by microscopy revealed that growth by coalescence was significant at saturations in the range 25–60%, above which the mass appeared overwetted.

Impeller rotation speed has an effect upon the granule growth of lactose, as shown in Fig. 13. It is to be expected that the agglomerate deformation produced by collision is dependent on the intensity of agitation, i.e. that the impeller rotation speed has an effect upon the agglomerate growth in addition to its effect upon the consolidation. Unpublished results have verified that the S–d_{gw} correlation is affected by the intensity of agitation when the starting materials are free flowing. Cohesive powders produce agglomerates with high strength because of the small particle sizes. The

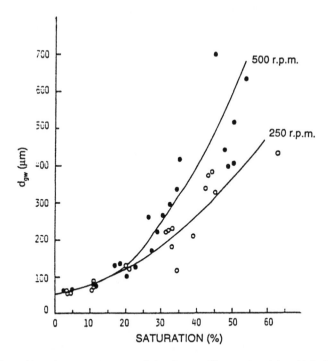

Fig. 13 Effect of liquid saturation in granulating lactose ($d_{gw} = 56\,\mu m$) in a high shear mixer. Impeller speed 250 r.p.m. (O) and 500 r.p.m. (●). Reproduced with permission from Kristensen *et al.* (1984), *Pharm. Ind.* **46**, 763–767. Editio Cantor, Denmark.

strength resists the stress produced by the impact so that the resulting strain becomes dependent only on the plasticizing effect of the liquid.

The fact that the particular lactose used in the study showed plasticity and growth by coalescence at low liquid saturations is probably partly due to dissolution of lactose in the binder liquid. In a study on melt granulation of lactose ($d_{gw} = 68 \, \mu m$) with polyethylene glycol (PEG) 3000 and 6000 in a 10-l Baker Perkins high shear mixer, Schaefer *et al.* (1990b) found that liquid saturations between 80 and 90% were necessary to achieve growth by coalescence (see section on process and product variables). With calcium hydrogen phosphate these authors found that significant growth by coalescence proceeded within the same range of liquid saturations. This result indicates that the error in the determination of the liquid saturations shown in Fig. 12 accounts for 10–15%, corresponding to an error of 2–3% in the determination of the granule density. Elema and Kristensen (1992) have recently compared the mercury immersion method for determining the density and porosity of pellets with a gas-permeametric method. They found that the results obtained by mercury immersion were systematically 10–15% higher than the results obtained by the gas-permeametric method.

The correlation between liquid saturation and granule size is valid in the liquid addition phase as well as the wet massing phase of agglomeration. This is due to the effect of liquid saturation on the strain behaviour of the agglomerates. The liquid saturation is controlled by the consolidation of the agglomerates and, thus, is particularly dependent on material properties such as particle size distribution, particle shape and surface texture. The correlation is assumed to be characteristic of a particular starting material or formulation insofar as the starting material has a particle size distribution which renders it cohesive. This is the case with most pharmaceutical formulations. The correlation may, therefore, be applied in analysing the effects of scaling up and comparisons between mixer–granulators.

Growth kinetics

Kapur (1978) presented an extensive review of the kinetics of wet granulation processes. On the supposition that the growth is controlled by a single mechanism, mathematical models for the changes in size and size distribution were presented. Such models are highly useful in analysing agglomeration processes, especially in the understanding of the agglomeration mechanisms and the resulting size distributions of a particular process. If, in the course of the agglomeration, successive granule size distributions exhibit similarity characteristics when plotted against an appropriate dimensionless size, it is reasonable to assume that there has been no change in the growth mechanisms governing the process. The cumulative granule size distribution is usually normalized by plotting it against d/d_{50}, where d_{50} is the median granule size. When the series of normalized distributions coincide, the distribution is said to be self-similar or self-preserving.

Published investigations on growth kinetics relate to low shear granulators such as rotating drums and there appear to be no studies in the pharmaceutical literature to establish growth kinetics. Leuenberger *et al.* (1990) presented a graph showing that wet granulation of lactose in a Diosna V10 high shear mixer produced a self-similar granule size distribution. In experiments with lactose–corn starch mixtures, they found that the resulting size distributions were similar to either the distribution obtained with lactose or the distribution for corn starch which means that increasing concentrations of corn starch at some point produce a change in the growth mechanism.

In agglomerate growth by nucleation it seems reasonable to postulate that the particles are well mixed, and that the collision frequency and probability of coalescence are independent of particle size. The agglomerate growth is then described by random coalescence kinetics. Kapur (1978) showed that these growth kinetics imply that the resulting size distributions are self-similar and that the rate of growth is described by the following relation:

$$V(t) = V_1 \exp[\lambda t/2] \tag{14}$$

where $V(t)$ and V_1 denote the mean agglomerate size at time t and the initial agglomerate size, respectively; λ is the specific coalescence rate which, when time-invariant, implies a straight-line relationship between t and log mean granule size.

Figure 14 shows the granule size distributions obtained by granulating a 8:2 mixture of lactose and corn starch with a 4% gelatine solution in a Glatt WSG 15 fluidized bed. The graph demonstrates self-similar granule size distributions which fit very well to a log-normal distribution with a geometric standard deviation of about 1.9. The addition of binder solution gave rise to a granule growth rate described by a straight line correlation between ln d_{50} and the added amount of binder solution, which is proportional to time t.

In their studies on fluidized bed granulation, Schaefer and Woerts (1978b) showed that the resulting granule size distributions are of log-normal type. The geometric standard deviation is almost constant in the course of the process but reduces when the added amount of liquid becomes relatively high. Other researchers, for example Ormos *et al.* (1975), have presented similar results. The growth mechanism is likely to change when a high proportion of granulating liquid has been added and all the primary particles are nucleated. In the course of the process the solvent evaporates turning the binder solution into a highly viscous, immobile liquid which

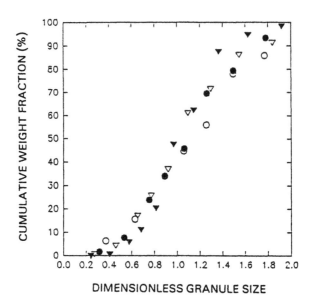

Fig. 14 Self-similar size distributions obtained by fluidized bed granulation of a lactose–corn starch mixture 8:2 with a 4% gelatine solution. Mass of added binder solution relative to mass of solids: 0.2, ○; 0.3, ●; 0.35, ▽; 0.45, ▼. Produced from data supplied from Schaefer and Woerts (1978b).

may facilitate further growth by coalescence of agglomerates and even balling because shear forces are absent.

Wet granulation in a high shear mixer proceeds by nucleation and coalescence mechanisms, the latter being dominant when the liquid saturation has been increased beyond a certain limit. In the case of calcium hydrogen phosphate, growth by nucleation is supposed to dominate until the liquid saturation has been increased to about 80% and above, c.f. Fig. 12.

Figure 15 demonstrates self-similarity of the size distributions achieved in the wet massing phase of granulating calcium hydrogen phosphate ($d_{gw} = 8.5\ \mu$m) in a high shear mixer. It is apparent that in the wet massing phase there is a change in size distribution from 3 to 6 min. The difference between the two distributions can be seen in context with the changing liquid saturation in the range of 55–80% at 0 and 3 min and above 80% at 6 and 8 min wet massing. The graph reflects, therefore, the change in growth mechanism from nucleation (0 to 3 min) into coalescence (6 to 8 min). Experiments with binder solutions of Kollidon 90, 25 and VA64 in varying concentrations show self-similar size distributions identical with the distributions shown in Fig. 15 and with the same distinction between

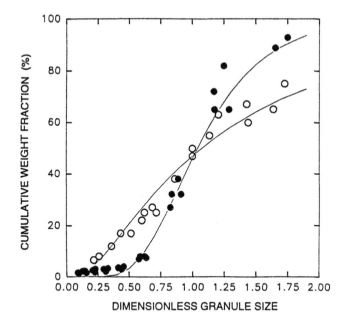

Fig. 15 Self-similar size distributions obtained by wet granulation in a Fielder PMAT 25 of calcium hydrogen phosphate with 10% m/m solution of hydrolysed gelatine. Wet massing times: 0 and 3 min, $\circ$; 6 and 8 min, $\bullet$. Solid lines: log-normal distributions with s_g = 2.01 and 1.48. Data from Holm *et al.* (1993).

the two stages according to the level of the liquid saturation.

The solid lines drawn in Fig. 15 are the estimated log-normal distributions. The geometric standard deviation s_g of the agglomerates in the nucleation stage of the process is about 2. The size distributions develop in parallel in the log-probability plot. When the growth mechanism changes into coalescence of agglomerates, the geometric standard deviation should change to value 1.5 and, at the same time, the growth rate should increase. In experimental work, higher values of s_g will normally be observed because, as shown in Fig. 16, the experimental size distributions over-represent larger agglomerates.

Nucleation of particles and coalescence of agglomerates involve similar mechanisms, as depicted in Fig. 3. Figure 15 demonstrates, however, that the size distribution achieved by the nucleation mechanism is wider than that achieved by coalescence of agglomerates. Growth by nucleation is by the kinetics of random coalescence which means that the collision frequency and the probability of coalescence are independent of size (Kapur, 1978). In contrast, growth by coalescence of agglomerates proceeds by the kinetics

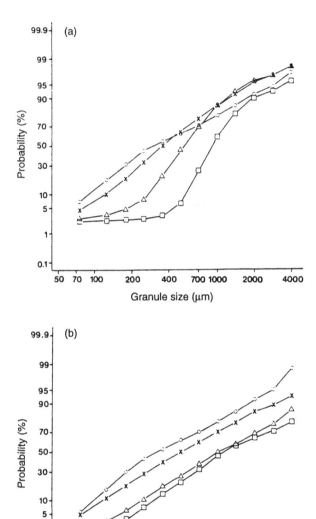

Fig. 16 Log-probability plot of the cumulative weight distribution during wet massing in Lödige FM 50. Granulation of calcium hydrogen phosphate with a 15% solution of Kollidon VA64 in water. a, Impeller speed 200 r.p.m., chopper speed 3000 r.p.m. b, Impeller speed 200 r.p.m., no chopper action. Wet massing times: 0 min, (O); 1 min, (x); 3 min, △; 6 min, □. Reproduced with permission from Schaefer et al. (1987), Pharm. Ind. **49**, 297–304. Editio Cantor, Denmark.

of non-random coalescence where the probability of coalescence is dependent on size as described earlier. It is a consequence of the non-random kinetics that a graph of the mean granule size against time is a straight line in a log–log plot (Kapur, 1978).

Figure 16 shows the granule size distributions obtained by granulating calcium hydrogen phosphate ($d_{gw} = 21 \mu$m) in a Lödige FM 50 mixer, which is a high shear mixer of horizontal type. It appears that the action of the chopper has a great influence on the granule size distribution. With the chopper inactive, the resulting size distributions are log-normal in distribution. The size changes are characterized by self-similar distributions and a geometric standard deviation of about 2.8. Schaefer *et al.* (1987) showed that the intragranular porosity remained almost constant (about 29%) during wêt massing. This means that the liquid saturation was also constant. With the chopper active, the intragranular porosity was reduced from 29 to 19% in the course of the process. The agglomerates must, therefore, have been saturated with liquid, which gives rise to growth by coalescence of agglomerates. Accordingly, Fig. 16 shows a change at 3 and 6 min into narrower size distributions. The straight-line part of the two distributions is described by a geometric standard deviation of about 1.7. The s_g of the entire distribution is greater (2.18 and 2.02, respectively) because the distributions are tailed upwards.

It is the author's experience that the granule size distributions shown in Fig. 16 are typical for wet granulation of cohesive powders in high shear mixers. When the nucleation mechanism dominates, the resulting granules have a wide size distribution, and the growth rate is described by a straight-line relationship between time and log mean granule size. When growth by coalescence of agglomerates becomes the dominating mechanism, the size distribution of the agglomerates is reduced because the probability of coalescence becomes size dependent (high for small agglomerates). The growth rate is then described by a straight-line relationship between time and mean granule size in a log–log plot.

Pelletization by wet granulation

Pelletization is a size enlargement process by which fine powders are converted into uniformly sized granules, preferably of spherical shape. Pellets for pharmaceutical purposes range in size, typically, from 0.5 to 1.5 mm, and are produced primarily for oral dosage forms with gastro-resistant or prolonged release properties. As drug delivery systems become more sophisticated, the role of pellets in the design and development of dosage forms is increasing.

Methods commonly used for pelletization of drug formulations are layering of drug substances from liquids or powders on inert spheres in coating pans or fluidized bed granulators, and extrusion/spheronization of moist, plasticized masses (Ghebre-Selassie, 1989). Wet-granulation in fluidized beds of the rotary type is a relatively new technique which seems to offer a great potential for pelletization purposes (Goodhart, 1989). Little attention has been paid to pelletization by wet granulation techniques although there is the possibility of a simple, one-step process with a short processing time in high shear mixers (Holm, 1987; D'Alonzo *et al.*, 1990; Zhang *et al.*, 1990).

Holm (1987) showed that wet granulation in a high shear mixer may result in narrowly sized and rounded granules provided that a high energy input is applied and that the amount of binder liquid is carefully controlled. He used a Fielder PMAT 25 mixer equipped with an impeller which allowed the blade angle to be varied. A high blade angle (40°) gave rise to a high power consumption and, hence, a high energy input resulting in rounded

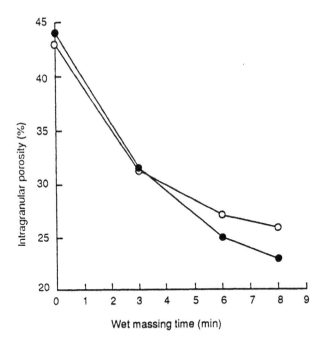

Fig. 17 The intragranular porosity of calcium hydrogen phosphate granulated with a 10% m/m solution of hydrolysed gelatine in water in a Fielder PMAT 25 high shear mixer; impeller blade angle 30° and impeller speed 400 r.p.m. Temperature of heating jacket: 12°C, ○; 45°C, ● . Reproduced with permission from Holm *et al.* (1993), *STP Pharma. Science* **3**, 286–293. Editions de Sante, France.

granules of uniform size. In order to ensure a uniform liquid distribution in the agitated mass, the walls of the mixer bowl were covered by a polytetrafluoroethylene (PTFE) tape in order to avoid deposition of the moistened powder.

Figure 15 shows that granulation of calcium hydrogen phosphate should result in a product with a geometric standard deviation of about 1.5. Experimental values are, however, greater because of the presence of larger granules. As growth by coalescence is size dependent, it should be possible to reduce the content of the larger agglomerates by reducing the amount of free surface liquid on the agglomerates, thereby delaying the growth of larger agglomerates. Holm *et al.* (1993) have recently shown that this can be done by drying in the wet massing stage. In this work, air was blown through the agitated mass, but it is likely that the same result can be achieved by vacuum techniques.

Figure 17 shows the change in granule porosity during wet massing in two series of experiments performed with cooling and heating, respectively, of the mantle of the mixer bowl. The graph demonstrates that the intragranular porosity of the agglomerates is affected by the product temperature. It shows that less liquid is required to achieve growth by the coalescence mechanism when the mantle is heated and, consequently, more liquid has to be removed by drying in order to delay the growth of larger agglomerates.

Figure 18 shows that the rate of removal of water from the product in the wet massing phase influences the geometric standard deviation. Holm *et al.* (1993) calculated the rate at which free liquid was supplied to the agglomerate surfaces because of densification. In experiments with a cooled mantle, the rate was about $10\,g\,min^{-1}$, while in the experiment with a heated mantle, the rate was $40\,g\,min^{-1}$. The calculations compare well with Fig. 18 showing that the most narrow size distributions were obtained when water was removed at a rate equal to that derived from densifying the agglomerates. The effect of the product temperature on the size dispersion is in agreement with equation 12 which predicts that the compressive strength of the denser agglomerates counteracts the growth rate. The resulting granules, with a geometric standard deviation of about 1.5, were approximately spherical.

The results shown in Fig. 18 describe the pelletization of a cohesive powder without addition of plasticizing agents. The plasticizing effect required to round the agglomerates is supplied entirely by the free surface liquid and supported by the high intensity of agitation. It is a prerequisite to the process that the rate by which the solvent is removed is adjusted carefully to the rate by which it is forced to the agglomerate surface by consolidation.

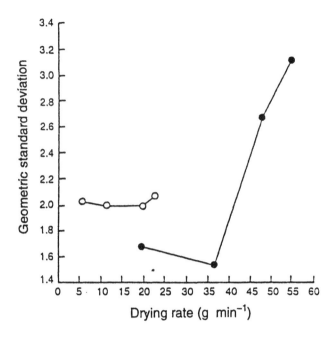

Fig. 18 Effect of drying rate upon the geometric standard deviation obtained in the same experiments as described in Fig. 17. Reproduced with permission from Holm et àl. (1993), *STP Pharma. Science* **3**, 286–293. Editions de Sante, France.

LIQUID REQUIREMENTS IN WET GRANULATION

Factors influencing the liquid requirements

It has been shown how the liquid requirements in wet granulation by mixer-granulators are determined by the correlation between liquid saturation and mean granule size. The liquid required to achieve a reproducible granulation by a particular process is, however, dependent on a range of factors related to the feed material as well as the mixer and its operation influencing the densification of the agglomerates and, hence, the liquid saturation. Linkson *et al.* (1973) examined published data on wet granulation of insoluble materials, and found that granulation, in general, required 50–55% v/v liquid. This result is probably valid for low shear mixers such as rotating drums, while wet granulation in high shear mixers requires less liquid because of a more pronounced densification.

It might be expected that the amount of liquid required to agglomerate a powder should equal the amount required to saturate the agglomerates (Capes *et al.*, 1977). This does not apply to all materials because moist agglomerates may become highly deformable at lower liquid saturations when particle interactions are weak, as demonstrated in experiments with lactose (Kristensen *et al.*, 1984), and with glass spheres which agglomerate at liquid saturations below 10% because of the smooth surface of the spheres (Holm *et al.*, 1985a).

It is likely that the use of relatively small amounts of granulating liquid, as applied in micro-granulation may suffice to remove the primary particles of the feed materials by nucleation. Das and Jarowski (1979) showed that micro-granulation leads to small granules with a relatively narrow size distribution. Moisture-activated dry granulation is a similar process which consists of an agglomeration stage that proceeds by the addition of small amounts of water to the dry solids containing a binder and, after agglomeration by agitation, the addition of a moisture-absorbing material, e.g. microcrystalline cellulose (Ullah *et al.*, 1987; Chen *et al.*, 1990). The resulting product is a dry granulation.

Early work has shown that the liquid requirements are strongly dependent on feed material properties and the type of mixer-granulator (Ganderton and Hunter, 1971; Hunter and Ganderton, 1972, 1973). With high shear mixers in particular, it is difficult to predict the liquid requirements within the narrow limits required to control and reproduce the process. For production purposes, the only practical approach is to employ instrumentation capable of detecting the phases of the process and, thus, the proper amount of binder liquid and the appropriate wet massing time (the granulation endpoint) to produce a granulation with the desired quality (size distribution, density, friability etc).

In fluidized bed granulation, granule growth is determined primarily by the moisture content of the bed and the droplet size of the atomized binder solution. The process proceeds by simultaneous liquid addition and solvent evaporation. If the moisture content is too high, the bed becomes overwetted and defluidizes rapidly. If the moisture content is too low, no agglomeration occurs. The liquid requirements are, therefore, determined primarily by the process conditions, especially by the liquid addition rate and the temperature and flow rate of the fluidizing air. The humidity of the air affects, to some extent, the moisture content of the bed.

A basic instrumentation for fluidized bed granulation must include the parameters determining the moisture content of the bed and the atomization of the binder solution, i.e. the inlet air temperature and humidity, inlet air-flow rate, liquid flow rate, and the pressure and flow rate of the atomizing air (Aulton and Banks, 1981; Kristensen and Schaefer, 1987).

254

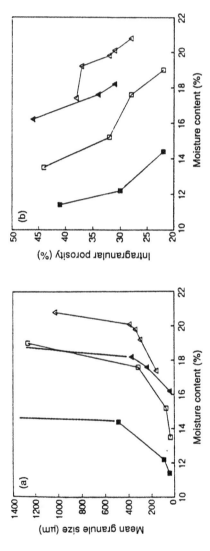

Fig. 19 Effect of moisture content on granule size (a) and intragranular porosity (b) in granulation of calcium hydrogen phosphate in two different mixer-granulators. Fielder PMAT 25: feed 7 kg, liquid addition rate 100 g min^{-1}, impeller speed 500 r.p.m. (■) and 250 r.p.m. (□). Lödige M5G: feed 1.5 kg, liquid addition rate 25 g min^{-1}, impeller speed 250 r.p.m. (▲) and 100 r.p.m. (△). Adapted from Schaefer et al. (1986a) with permission from the authors.

Comparison between mixer-granulators

Figure 19 shows the effect of moisture content upon mean granule size when granulating calcium hydrogen phosphate with a PVP–PVA copolymer solution. With the Fielder mixer, considerably less liquid is required to achieve a certain granule size than with the Lödige mixer. In addition, the impeller rotation speed affects the liquid requirements. Examination of the porosity changes during the process shows that the Fielder mixer is more efficient at densifying the agglomerates, and that the resulting low porosity is associated with a corresponding low liquid requirement. The effects of the different liquid requirements and porosities are cancelled out by plotting granule size against liquid saturation. The resulting graph coincides with the correlation shown in Fig. 12.

When the Fielder mixer was operated with an impeller speed of 500 r.p.m., the agglomerate porosity was reduced to about 20% which gave rise to overwetting of the mass and, consequently, uncontrolled growth, as indicated by the dotted lines. This demonstrates the potential risk for 'overgranulation' or overwetting by wet granulation in high shear mixers of cohesive powders which consolidate steadily during the process. A change in porosity of, say, 2% produces a change in liquid saturation of about 10%, c.f. equation 13, which in the stage of rapid growth by the coalescence mechanism may turn the process into uncontrolled growth.

Schaefer et al. (1986b, 1987) compared different high shear mixers available in the Danish pharmaceutical industry. The comparison was based on wet granulation of calcium hydrogen phosphate. The different growth rates obtained in the different mixers could be attributed to differences in intensity of agitation, as expressed by the swept volume. Surprisingly, scaling up from laboratory to production scale mixers had only a minor effect on the liquid requirements.

Richardson (1982) characterized the impellers in different high shear mixers by their relative swept volumes. The vertical volume swept out by the impeller blades at each revolution is calculated by dividing the blade area into vertical segments. On the basis of this volume and the impeller speed, the volume swept out per second can be calculated and divided by the volume of the mixer bowl in order to obtain the relative swept volume. The relative swept volume thus depends on the vertical dimensions of the impeller blade, on the impeller speed, and on the size of the bowl. The relative swept volume is a measure of the energy input of the impeller blades on the material. Accordingly, a higher relative swept volume was found to result in a greater increase of the product temperature and in denser granules when comparing high shear mixers of different type (Schaefer, 1988).

Table 2 compares Diosna and Fielder mixers of different scale. In the

Table 2 The relative volume swept out by the impeller in Fielder and Diosna mixers of different scale (Schaefer, 1988)

	r.p.m.[a]	Relative swept volume per second[a]
Diosna P 25	150/300	1.37/2.74
Diosna P 50	144/188	1.08/2.16
Diosna P 250	95/190	0.52/1.03
Diosna P 400	65/130	0.36/0.72
Fielder PMAT 25	150/300	0.75/1.51
Fielder PMAT 65	150/300	0.71/1.42
Fielder PMAT 150	127/254	0.61/1.23

[a] Low/high impeller rotation speed.

Diosna mixers, scaling up results in a marked decrease in the relative swept volume and consequently in a lower power input. As a consequence, the resulting granule porosity increases in scaling up. Richardson (1982) showed that the fall in relative swept volume can be compensated for by a longer wet massing time in production-scale mixers, thereby obtaining denser granules. An alternative is to change the relative swept volume by modifying the impeller design or by variation of the impeller speed. The latter method is questionable if it results in a significant difference in the centrifugal forces in the different mixers.

Table 2 shows that there is only a slight decrease in relative swept volume when scaling-up in Fielder mixers. In laboratory and pilot scales the relative swept volume is higher in Diosna mixers than Fielder mixers. Accordingly, Schaefer et al. (1987) found that the Diosna mixers produced denser granules than did the Fielder mixers. They observed no significant differences between the Diosna P 250 and Fielder PMAT 150 where the swept volumes are similar.

Direct scaling up of the amount of binder solution by adding the same relative amount on different scales does not lead to a constant moisture content in the mass. The marked heat production, caused by the high power input in high shear mixers, results in the evaporation of water which is pronounced in the small mixers because of the more intensive agitation. In extreme cases, the loss of water may account for more than 15% of the amount added (Kristensen, 1988a). In the study by Schaefer et al. (1986b, 1987), this meant that approximately the same amount of binder liquid was required in small- and large-scale mixers. When scaled up the reduced evaporation of water compensated for the higher intragranular porosity so that the liquid saturation level was kept almost constant.

An attempt to identify the parameters of scaling up has recently been presented by Horsthuis *et al.* (1993) who compared wet granulation of lactose in three high shear mixers (Gral 10, 75 and 300). The three mixers are not geometrically similar. The relative swept volume is, therefore, strongly dependent on the mixer scale. It was found that the granulation process with respect to temperature increase and granule size distribution could be scaled up by keeping the Froude number constant. The dimensionless Froude number expresses the value of N^2D/G where N is the revolutions per minute, D is the diameter of the impeller and G is the gravitation constant. The Froude number is the ratio of the centrifugal force to the gravitational force. In their experiments, a constant relative swept volume or a constant impeller tip speed did not result in a comparable process. This particular study demonstrates that the relative swept volume is insufficient to predict scaling up when there is a large difference in scale between the different mixers. The centrifugal forces should also be kept approximately constant in addition to the relative swept volume in order to achieve comparable processes in the scaling up. A variation in centrifugal forces means that the part of the impeller blades which is effective in wet massing also varies.

Starting material properties

Variations in starting material properties may have a significant effect upon the amount of liquid and the wet-massing time required to reproduce the granulation. Except for the effect of the particle size distribution, little systematic work is found in the literature on effects of the starting material properties. It can, however, be inferred from the agglomeration mechanisms that the following material properties influence granule formation and growth:

1. Wetting and spreading of the binder solution on the solid;
2. Solubility of the solid in the binder solution;
3. Size and size distribution of the solid;
4. Particle shape and surface morphology;
5. The packing properties of the solids.

It is a general experience that the smaller the particle size of the starting material, the more binder liquid is required to achieve agglomeration; previously this was attributed to an effect of the specific surface of the feed. Taking into consideration the correlation between liquid saturation and mean granule size, it might be more fruitful to consider the growing liquid requirements as an effect of the packing properties of the particle system.

Particle size, size distribution, shape and surface morphology all affect the packing density and the ease by which dense packings are achieved. The packing properties affect the intragranular porosity and, hence, the liquid saturation.

Kristensen *et al*. (1985a,b) investigated the agglomeration of different size fractions of calcium hydrogen phosphate. They found that, independent of the particle size, liquid saturations close to 100% are required to render the moist agglomerates plastically deformable. Growth by the coalescence mechanism is, therefore, influenced primarily by the tensile strength of the agglomerates. According to equation 5 the tensile strength is proportional to the specific surface area of the solid because, in the case of a polydisperse powder, the volume–surface diameter d_{vs} is substituted for d in the equation. Thus, granulation experiments in a Fielder PMAT 25 high shear mixer showed proportionality between the liquid requirement and the compressive strength of agglomerates. The liquid saturation required to see growth by coalescence was strongly dependent on particle size.

The effect of lactose particle size was investigated by Schaefer *et al*. (1990a) in a study on wet granulation in a 10-1 Baker Perkins high shear mixer. They found, in agreement with other workers (Lindberg *et al*., 1985; Paris and Stamm, 1985; Tapper and Lindberg, 1986), that as the particle size decreased an increasing amount of binder solution was needed to achieve a particular granule size. In contrast to the results obtained with different particle sizes of calcium hydrogen phosphate, Schaefer *et al*. (1990a) found that the correlation between liquid saturation and mean granule size was independent of the particle size of lactose, and that the different liquid requirements, therefore, were due to different granule porosities. In no case did the liquid saturation exceed about 70%. Tapper and Lindberg (1986) found that a coarse lactose agglomerated at lower liquid saturations than a fine material. The conflicting results may be due to the use of different mixer-granulators and, possibly, a bias in Schaefer *et al*.'s determination of the granule density and porosity caused by a size-dependent dissolution of lactose in the binder liquid.

Although there are many reports in the literature on liquid requirements to granulate pharmaceutical formulations, a simple relationship between the liquid requirements, the properties of the formulations and the agglomerate growth is difficult to establish. In many cases, the experimental procedure involves milling and sieving of the dried granulation before determination of the size distribution. This further affects the characteristics of the agglomerates. Other investigations, such as recent work by Wehrle *et al*. (1992), Vojnovic *et al*. (1992) and Shirakura *et al*. (1992), employ experimental designs of multivariate type to produce evidence about the factors critical to the process but not necessarily detailed evidence about effects of each single factor.

The effect on granule growth during wet granulation of adding corn starch to calcium hydrogen phosphate has been reported by Holm *et al.* (1985b) for wet granulation in a Fielder PMAT 25 mixer and by Schaefer *et al.* (1990a) for granulation in a 10-l Baker Perkins mixer. In both cases, it was found that the addition of corn starch improved the rate of consolidation so that the minimum intragranular porosity (30–35%) was achieved early in the wet massing phase.

The robustness of a wet granulation process against the inevitable variations in raw material properties, especially variations in particle size and size distribution, is clearly dependent on the changes in agglomerate porosity occuring during the process. Starting materials, which have the characteristics of cohesive powders, are likely to produce agglomerates that consolidate steadily during the entire process by a rate and extent dependent on the energy input by the agitators. The liquid requirement and the granule growth are, therefore, strongly dependent on the type of mixer-granulator and its operation as shown in Fig. 19. Addition of starches and similar materials to the cohesive feed may improve the robustness of the process because they counteract densification of the agglomerates. Free-flowing starting materials are likely to produce agglomerates that are consolidated to their minimum porosity early in the process. Because of the ease of consolidation, the liquid requirement and final granule size are relatively insensitive to a change of mixer and its operation. Once the minimum agglomerate porosity has been achieved, loss of liquid by evaporation due to heat produced by the agitation is the primary factor affecting the final granule size.

Control of wet granulation

It is outside the scope of this chapter to discuss production aspects of agglomeration processes. The readers are referred to literature on instrumentation for the control of wet granulation, for example the review by Kristensen and Schaefer (1987) which includes a chapter on end-point control of wet massing methods and Werani (1988) showing the benefits of power consumption measurements for production purposes. However, a few comments on end-point control are given in order to illustrate their relation to agglomeration mechanisms.

Leuenberger *et al.* (1979) investigated the liquid requirements during granulation in low shear mixers and provided an equation expressing the optimum amount of liquid as a function of granule porosity, feed equilibrium humidity and density of feed and liquid. Experiments on lactose–corn starch mixtures demonstrated that the liquid requirement can be predicted from measurements of the tapped density of the dry feed material (Bier *et al.*, 1979; Leuenberger *et al.*, 1979; Leuenberger, 1982). The predicted amount of liquid agreed well with granulation experiments in planetary and Z-blade mixers.

A variety of instrumentation has been proposed for control of liquid volume and process time. Such instrumentation should measure the changes in the rheological properties of the mass. The potential of torque measurements for detecting the stages of wet granulation processes has been well documented, as has the measurement of the bending moment on a probe inserted into the agitated mass. A simple mixer torque rheometer has recently been proposed by Hancock *et al.* (1991) to study the rheological behaviour of moistened masses and to assess binder–solid interactions. A similar idea was presented by Alleva and Schwartz (1986).

Figure 20 shows a power consumption profile recorded during addition of a povidone solution to a lactose–corn starch mixture in a planetary mixer. Similar profiles have been recorded for wet granulation in high shear mixers (Leuenberger *et al.*, 1990). The profile reflects the characteristic phases of the agglomeration process (Leuenberger, 1982). In phase I, the powder is dry mixed and the amount of liquid is insufficient to create pendular liquid bondings. The power consumption remains unchanged. At a certain point, phase II, a sharp increase in power consumption is seen

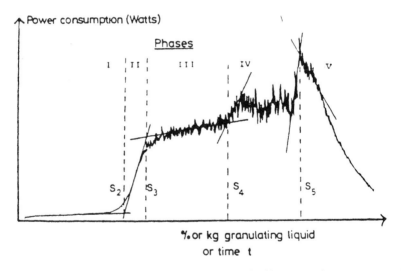

Fig. 20 Power consumption profile recorded during liquid addition to a lactose–corn starch mixture in a planetary mixer. Reproduced from Leuenberger *et al.* (1979) with permission from the authors.

because the amount of liquid now suffices to establish liquid bondings. In phase III, the power consumption profile levels off while the liquid phase fills up the intragranular voids and coarser granules are produced. In phase IV, the granules are filled with liquid. The power consumption drops when reaching S_5 which corresponds to 100% saturation.

Leuenberger *et al*. (1979), using different scale planetary mixers, showed that the estimated values of S_3 and S_4 are proportional to the batch size and that the liquid required to run a robust process lies in the range S_3–S_4. The amount of applied liquid can be normalized as follows (Leuenberger *et al*., 1990):

$$\pi = \frac{S - S_2}{S_5 - S_2} \tag{15}$$

where π is the normalized liquid quantity, S is the absolute liquid quantity, and S_5 and S_2 are parameters of the power consumption profile. The authors stated that π is equivalent to the liquid saturation of the moist agglomerates, and that the use of π enables a direct comparison of the agglomeration properties of different starting materials. This agrees very well with the effect of liquid saturation on granule growth described earlier.

The energy consumption by wet granulation, i.e. the cumulated power consumption profile, is converted completely into heating of the mixer and its content (Holm *et al*., 1985a). This is in agreement with the earlier discussion of collision and coalescence of particles. The absorption of the kinetic energy of the particles results in heat. When growth by coalescence of agglomerates becomes significant, because of high agglomerate deformability, the energy consumption will increase accordingly. There is, therefore, a fundamental correlation between the power consumption profile and granule growth because both are influenced by the agglomerate deformability. The correlation between power consumption and granule growth is demonstrated in several papers, see for example Holm *et al*. (1985a,b) and Werani (1988).

It appears that it is the change in liquid saturation during wet massing which determines the growth kinetics and, thus, the growth rate and the energy consumption of the moist mass. In a particular process, the only change affecting the liquid saturation in the wet massing stage is the densification of the agglomerates. This means that there should be a correlation between the porosity function $(1 - \epsilon)/\epsilon$, which determines the changing liquid bonding strength, c.f. equation 5, and the power consumption as well as the mean granule size. This was verified by Ritala *et al*. (1988) in a study on granulation of calcium hydrogen phosphate with various binder solutions in a high shear mixer. The authors compared the data obtained in the late stages of the process where the agglomerates can be presumed to be saturated with liquid.

MELT GRANULATION

Process and equipment

Melt granulation – also called thermoplastic granulation – is an agglomeration process based upon the use of a binder material which is solid at room temperature and softens or melts at slightly elevated temperatures, usually above 50–60°C. When melted, the action of the liquid binder is similar to the action of a binder solution in a wet granulation process. The binders normally used for melt granulation are polyethylene glycols (Rubinstein, 1976; Ford and Rubinstein, 1980; Pataki et al., 1983; Kinget and Kemel, 1985; Schaefer et al., 1990b). The use of hydrophobic binders such as waxes and stearic acid have been investigated for the purpose of preparing sustained release products (McTaggert et al., 1984; Flanders et al., 1987). A range of hydrophobic, meltable substances for the preparation of matrix pellets with prolonged release properties have been investigated by Thomsen et al. (1993b).

When using melt granulation the stages of liquid addition and drying of a wet granulation process are eliminated. It is, perhaps, of greater importance that granulation of water-sensitive materials is possible. A disadvantage for heat-sensitive materials is the elevated product temperature required to ensure melting and distribution of the binder material. In some processes, the final product temperature may be increased to more than 100°C and, thus, cause liberation of water of crystallization.

The binder material is added to the starting materials either as a powder or in molten form to the preheated starting materials. The product must be heated to a temperature above the melting point of the binder. This can be achieved by a heating jacket (Kinget and Kemel, 1985) or by heat of friction caused by intensive agitation (Schaefer et al. 1990b). At a production scale, it is advantageous to use a high shear mixer with a power input sufficient to generate the required product temperature within an acceptable time. Figure 21 compares the correlation between relative swept volume and impeller rotation speed in different laboratory-scale high shear mixers. The Baker Perkins 10 mixer is clearly capable of producing a high energy input. Flanders et al. (1987) examined three scales of Baker Perkins high shear mixers (10, 60 and 600 l) and found that melt granulation by heat produced solely by friction is possible in all of them. The authors mentioned that the Fielder and Diosna high shear mixers are unsuitable for melt granulation

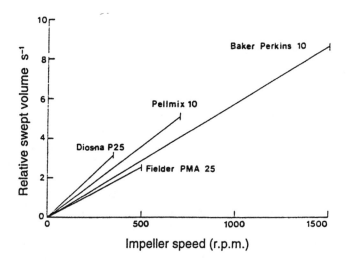

Fig. 21 Correlation between relative swept volume and impeller rotation speed in laboratory scale high shear mixers. Reproduced from Schaefer *et al.* (1992a) with permission from the authors.

purposes because of the long process time required to achieve the necessary product temperature. The Pellmix 10 high shear mixer, which has a bowl volume of 50 l, is shown to possess the necessary energy input to ensure melting by frictional heating (Schaefer *et al.*, 1992a). In the work by Kinget and Kemel (1985), a Gral 10-1 mixer equipped with a heating jacket was used. The molten binder was added to the preheated powder.

When melt granulation is performed in a high shear mixer that provides a high energy input to the product, the physical conditions required to produce rounded granules with a narrow size distribution, i.e. pellets, are met. Melt granulation in high shear mixers has, therefore, potential as a simple and fast method for pelletization (see later).

Process and product variables

Schaefer *et al.* (1990b) investigated melt granulation in a 10-1 Baker Perkins high shear mixer using polyethylene glycol (PEG) 3000 and 6000. The starting materials were calcium hydrogen phosphate ($d_{gw} = 23\ \mu$m) and lactose ($d_{gw} = 68\ \mu$m) of a quality identical to that used in a study on wet granulation in the same mixer (Schaefer *et al.*, 1990a). Direct comparisons between wet and melt granulation can therefore be made. A conclusion of the work, which agrees well with the conclusions made by Kinget and Kemel (1985) in their study on melt granulation in a 10-l Gral mixer, is that the main factors influencing agglomeration are the relative amount of binder and its viscosity, the impeller rotation speed and massing time. Except for an effect of liquid binder viscosity, the effects of the other factors mentioned agree well with results of wet granulation.

Schaefer et al. (1990b) showed that the agglomerate growth of calcium hydrogen phosphate can be correlated with the liquid saturation of the moist agglomerates. Rapid growth by the coalescence mechanism was seen at 80–85% saturation while a slightly higher saturation was observed by wet granulation. As discussed earlier, the difference is supposed to be caused by an error in the measurement of the density of wet granulated granules (Elema and Kristensen, 1992). Granules prepared by melt granulation contain the solidified binder and are, therefore, less porous than wet granulated granules. Smaller surface pores give rise to reduced penetration by mercury during the measurement. The optimum amount of binder to agglomerate the particular calcium hydrogen phosphate was in the range of 37–43% v/v at melt granulation and slightly higher by wet granulation, because loss of water by evaporation affects the liquid requirements during wet granulation in high shear mixers.

The results obtained with lactose were different. Agglomerates of lactose were consolidated to their minimum porosity after a short massing time. The ease of consolidation is affected by the particle size of the feed material. In wet granulation processes this means that the granule size remains unchanged during further massing because the liquid saturation is constant. In contrast, melt granulation of lactose showed a constant growth with massing time, despite constant saturation with the liquid binder: the higher the content of binder, the higher the growth rate. The amount of binder necessary for wet granulation (14.4% v/v) was considerably less than that required for melt granulation (21.2–25.4% v/v). The range of liquid saturations required to produce growth by coalescence by melt granulation was 80–90%, i.e. the same range as determined for calcium hydrogen phosphate. The different liquid requirements for agglomeration of lactose are undoubtedly caused by dissolution of lactose in the aqueous binder liquid.

Figure 22 shows the granule growth of lactose (α-monohydrate lactose) by melt granulation in a high shear mixer. Except for the effect of impeller rotation speed, which gives rise to different mean granule sizes, the graph demonstrates that the use of PEG 3000, independently of impeller speed, produces larger granules than does PEG 6000. Liquified polyethylene glycols wet lactose and have the same surface tension. The viscosity of PEG 3000 at 90°C is about 135 mPa s, and that of PEG 6000 about 500 mPa s – the less viscous binder liquid produces the largest granules. If the effect of

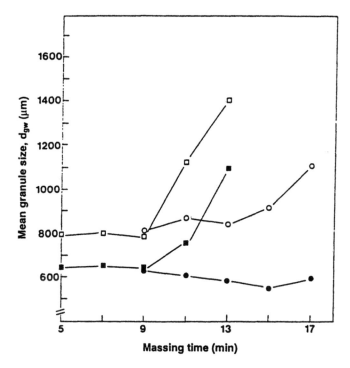

Fig. 22 Effects of impeller rotation speed and polyethylene glycol (PEG) grade upon granule growth by melt granulation of 450 mesh lactose in a Pellmix 10 high shear mixer. Binder concentration 23% v/v; PEG 3000 (O, □); PEG 6000 (●, ■); impeller speed 500 r.p.m. (O, ●) and 700 r.p.m. (□, ■). Reproduced from Schaefer *et al.* (1992a) with permission from the authors.

the two PEGs on the growth results from the different viscosities, it conflicts with the claimed effect of liquid viscosity upon the probability of coalescence of agglomerates. Unpublished results have, however, shown that the product temperature, varied by applying a heating jacket, has an effect upon the growth. The higher the temperature and, consequently, the lower the binder viscosity, the smaller the granules produced by experiments using one of the two PEG grades. The difference between the effects of PEG 3000 and 6000 on the growth may be attributed to the very high viscosity of the molten PEG 6000, or it may be caused by interactions between the liquid binder and the lactose. The effect of binder concentration upon growth of two types of lactose shown in Fig. 23 may also be attributed to binder–substrate interactions.

Figure 23 shows the effect of lactose particle size upon the mean granule

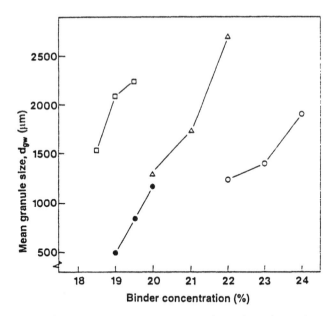

Fig. 23 Effect of binder concentration upon mean granule size by melt granulation of different lactose qualities in a Pellmix 10 high shear mixer. Impeller rotation speed 700 r.p.m.; 13 min massing time; α-monohydrate lactose 200 mesh (□), 350 mesh (△) and 450 mesh (○); anhydrous lactose (●). Reproduced from Schaefer *et al.* (1992c) with permission from the authors.

size by melt granulation of lactose with PEG 3000. The grade and particle size d_{gw} of the lactose were mesh 200 (44 μm), mesh 350 (34 μm), mesh 450 (22 μm) and anhydrous lactose (13 μm). The effects of particle size, binder concentration and additional results on the effect of impeller rotation speed upon the growth of lactose agglomerates are in good agreement with the results of wet granulation of the material. They can be attributed to the effect of the agglomerate consolidation during the process. It is surprising, however, that anhydrous lactose, which has the smallest mean particle size, is agglomerated by less liquid than a lactose of similar particle size, and that larger granules than those shown in the graph could not be produced.

Although there are similarities between melt and wet granulation as to the effects of intensity of agitation, process time, particle size of the starting material and liquid requirements, there appear to be additional factors to be taken into account. The effect of the PEG grade upon granule growth (Fig. 22) and the effect of feed material properties (Fig. 23) are examples of factors that do not influence wet granulation processes to the same extent.

Melt pelletization

High energy input, a controlled amount of liquid phase, and a uniform liquid distribution, are the prerequisites to produce pellets by wet granulation; accordingly, melt granulation in high shear mixers is likely to be appropriate for pelletization of powders insofar as the mixer can supply the required energy input. Schaefer *et al.* (1992a,b,c) have demonstrated that commercial qualities of calcium hydrogen phosphate and lactose can be pelletized with PEGs in a Pellmix 10 high shear mixer. The size distribution of the resulting granules had geometric standard deviations in the range 1.3 to 1.4, which compares well with the size distribution of pellets prepared by extrusion/spheronization. The resulting pellets may appear as smooth, rounded granules. Compared with wet granulation methods, melt pelletization has the advantage that the amount of binder material is easier to control because there is no evaporation during the process.

The amount of meltable binder material required to pelletize has to be adjusted to the particular feed material and to the actual process conditions. According to the work by Schaefer *et al.*, the variation of the binder concentration that can be tolerated is less than 1–2% of the added amount of PEG. Significant deposition of the moistened mass to the walls of the mixer must be avoided in order to ensure the homogeneity of the mass (Thomsen *et al.*, 1993a,b). In the Pellmix 10 mixer, this is achieved by a PTFE-lining of the walls of the mixer bowl. In a study on melt granulation in a 10-l Baker Perkins mixer, Schaefer *et al.* (1990b) found that PEGs are especially suitable for melt granulation. Other meltable binders (stearyl alcohol, stearic acid, cetyl alcohol, glycerol monostearate) gave rise to deposition of moistened mass to the walls of the mixer bowl. Investigation of melt granulation with hydrophobic binders in a laboratory-scale Pellmix PL 1/8 mixer has shown that the deposition of moist mass may be the determining factor in the choice of meltable binder (Thomsen *et al.*, 1993a, b).

Effects of process variables on melt pelletization of lactose using PEGs are reported by Schaefer *et al.* (1993a). The authors compared the process in two different scale mixers (Schaefer *et al.*, 1993b).

Melt pelletization in high shear mixers is a simple and rapid process compared with other methods for pelletization of powders and has the advantage of being solvent free. An additional benefit is that the pelletized product may contain up to about 85% m/m active ingredient. Melt pelletization using PEG as meltable binder renders the pellets hydrophilic. If a hydrophobic binder is used instead, the pellets may have prolonged release properties, as demonstrated by Thomsen (1994).

Figure 24 shows release profiles of pellets prepared by melt pelletization of paracetamol with a meltable binder composed of glycerol monostearate

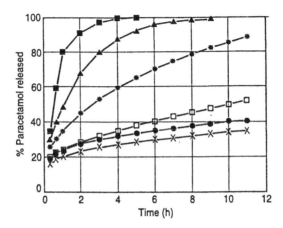

Fig. 24 Release profiles of paracetamol pellets prepared by melt pelletization using mixtures of glycerol monostearate (GMS) and microwax (MW) as meltable binder. Ratio of GMS to MW: 1:0, ■; 7:1, ▲, 3:1, *, 1:1, □, 1:3, ●; and 1:7, x. Reproduced from Thomsen (1992).

and microcrystalline wax (a paraffin). The rate of release of paracetamol can be varied within wide limits by changing the ratio of glycerol monostearate to the wax. The mechanism of release from the matrix of binder material is diffusion controlled. A clinical study (unpublished) has shown that the matrix of glycerol monostearate and microcrystalline wax resists the gastrointestinal fluids, and that there is a close agreement between the *in vivo* release determined from plasma concentration profiles and the *in vitro* release determined by the paddle apparatus described in the *European Pharmacopoeia* and using a phosphate buffer of pH 6.8 as the dissolution medium.

CONCLUDING COMMENTS

The fundamentals of particle agglomeration have been discussed. It has been shown that a fundamental parameter of agglomeration processes is the degree of liquid saturation of the agglomerates. The liquid saturation, being dependent on the content of liquid phase and the particle packing density of agglomerates, controls the growth kinetics and the resulting granule size and size distribution. The correlation between liquid saturation and mean granule size is specific for the feed material. It may be applied to compare granulation in mixer-granulators of different type and scale by comparing the efficiency of the mixer in consolidating the particular feed material.

It is apparent that wet and melt granulation processes can be controlled to produce granules with the desired quality attributes – size and size distribution, density, friability, etc. The direction of changes in granule growth induced by changing the granulating equipment and its operation and by varying the feed material properties can be controlled. It is still impossible, however, to predict the liquid requirements of a particular process. And very little is known about the optimum design of mixer-granulators and interactions between mixer design and feed material properties.

Future work on particle agglomeration processes should focus on the relations between solid–liquid interactions on the microscale and the kinetics of agglomerate formation and growth. By this route, it might be possible to develop regimens of particle agglomeration that allow direct comparisons between different types of mixers and different starting materials.

ACKNOWLEDGEMENTS

The Danish Technical Research Council and Pharmacia Therapeutics AB in Sweden are acknowledged for their funding of the research on particle agglomeration at the Department of Pharmaceutics, The Royal Danish School of Pharmacy. Associate Professor Torben Schaefer, PhD, is acknowledged for valuable help in preparing this monograph.

REFERENCES

Alleva, D. S. and Schwartz, J. B. (1986). *Drug Dev. Ind. Pharm.* **12**, 471–487.

Aulton, M. E. and Banks, M. (1981). *Int. J. Pharm. Tech. & Prod. Mfr.* **2** (4), 24–29.

Bier, H.-P., Leuenberger, H. and Sucker, H. B. (1979). *Pharm. Ind.* **41**, 375–380.

Bowden, F. P. and Tabor, D. (1964). *The Friction and Lubrication of Solids. Part I*, pp. 299–306. Oxford University Press, Oxford.

Capes, C. E. (1980). *Particle Size Enlargement*. Elsevier, Amsterdam.

Capes, C. E., Germain, R. J. and Coleman, R. D. (1977). *Ind. Eng. Chem., Process Des. Dev.* **16**, 517–518.

Chan, S. Y., Pilpel, N. and Cheng, D. C.-H. (1983). *Powder Technol.* **34**, 173–189.

Chen, C.M., Alli, D., Igga, M. R. and Czeisler, J. L. (1990). *Drug Dev. Ind. Pharm.* **16**, 379–394.

Cheng, D. C.-H. (1968). *Chem. Eng. Sci.* **23**, 1405–1420.

D'Alonzo, G. D., O'Connor, R. E. and Schwartz, J. B. (1990). *Drug Dev. Ind. Pharm.* **16**, 1931–1944.

Das, S. and Jarowski, C. I. (1979). *Drug Dev. Ind. Pharm.* **5**, 479–488.

Eaves, T. and Jones, T. M. (1971). *Rheol. Acta.* **10**, 127–134.

Eaves, T. and Jones, T. M. (1972a). *J. Pharm. Sci.* **61**, 342–348.

Eaves, T. and Jones, T. M. (1972b). *Pharm. Acta Helv.* **47**, 537–545.

Elema, M. O. and Kristensen, H. G. (1992). *Acta Pharm. Nord.* **4**, 233–238.

Ennis, B. J., Tardos, G. and Pfeffer, R. (1991). *Powder Technol.* **65**, 257–272.

Flanders, P., Dyer, G. A. and Jordan, G. (1987). *Drug. Dev. Ind. Pharm.* **13**, 1001–1022.

Fonner, D. E., Anderson, N. R. and Banker, G. S. (1982). In *Pharmaceutical Dosage Forms. Tablets Vol. 2* (Lieberman, H. A. and Lachman, L., eds), pp. 185–267. Dekker, New York.

Ford, J. L. and Rubinstein, M. H. (1980). *Pharm. Acta Helv.* **55**, 1–7.

Ganderton, D. and Hunter, B. M. (1971). *J. Pharm. Pharmacol.* **23** Suppl., 1S–10S.

Ghebre-Sellassie, I. (ed.) (1989). *Pharmaceutical Pelletization Technology*. Dekker, New York.

Goodhardt, F. W. (1989). In *Pharmaceutical Pelletization Technology* (Ghebre-Sellassie, I., ed.), pp. 101–122. Dekker, New York.

Hancoock, B. C., York, P., Rowe, R. C. and Parker, M. D. (1991). *Int. J. Pharm.* **76**, 239–245.

Ho, T. and Hersey, J. A. (1979). *Powder Technol.* **23**, 191–195.

Holm, P. (1987). *Drug Dev. Ind. Pharm.* **13**, 1675–1701.

Holm, P., Jungersen, O., Schaefer, T. and Kristensen, H. G. (1983). *Pharm. Ind.* **45**, 806–811.

Holm, P., Schaefer, T. and Kristensen, H. G. (1985a). *Powder Technol.* **43**, 213–223.

Holm, P., Schaefer, T. and Kristensen, H. G. (1985b). *Powder Technol.* **43**, 225–233.

Holm, P., Schaefer, T. and Kristensen, H. G. (1993). *STP Pharma Science* **3**, 286–293.

Horsthuis, G. J. B., van Laarhoven, J. A. H., van Rooij, R. C. B. M. and Vromans, H. (1993). *Int. J. Pharm.* **92**, 143–150.

Hunter, B. M. and Ganderton, D. (1972). *J. Pharm. Pharmacol.* **24**, Suppl., 17P–24P.

Hunter, B. M. and Ganderton, D. (1973). *J. Pharm. Pharmacol.* **25**, Suppl., 71P–78P.

Jaegerskou, A., Holm, P., Schaefer, T. and Kristensen, H. G. (1984). *Pharm. Ind.* **46**, 310–314.

Jäger, K.-F. and Bauer, K. H. (1984). *Acta Pharm. Technol.* **30**, 85–92.

Kapur, P. C. (1978). *Adv. Chem. Eng.* **10**, 55–123.

Kinget, R. and Kemel, R. (1985). *Acta Pharm. Technol.* **31**, 57–62.

Kristensen, H. G. (1988a). *Acta Pharm. Suec.* **25**, 187–204.

Kristensen, H. G. (1988b). Binders. In *Encyclopedia of Pharmaceutical Technology* (Swarbrick, J. and Boylan, J. C., eds), Vol. 1, pp. 451–464. Dekker, New York.

Kristensen, H. G. and Schaefer, T. (1987). *Drug Dev. Ind. Pharm.* **13**, 803–872.

Kristensen, H. G. and Schaefer, T. (1992). Granulations. In *Encyclopedia of Pharmaceutical Technology* (Swarbrick, J. and Boylan, J. C., eds), Vol. 7, pp. 121–160. Dekker, New York.

Kristensen, H. G., Jaegerskou, A., Holm, P. and Schaefer, T. (1984). *Pharm. Ind.* **46**, 763-767.

Kristensen, H. G., Holm, P. and Schaefer, T. (1985a). *Powder Technol.* **44**, 227-237.

Kristensen, H. G., Holm, P. and Schaefer, T. (1985b). *Powder Technol.* **44**, 239-247.

Krycer, I., Pope, D. G. and Hersey, J. A. (1983). *Powder Technol.* **34**, 30-51.

Leuenberger, H. (1982). *Pharm. Acta Helv.* **57**, 72-82.

Leuenberger, H., Bier, H.-P. and Sucker, H. B. (1979). *Pharm. Technol. Int.* **2** (3), 35-42.

Leuenberger, H., Luy, B. and Studer, J. (1990). *STP Pharma Sciences* **6**, 303-309.

Linkson, P. B., Glastonbury, J. R. and Duffy, G. J. (1973). *Trans. Inst. Chem. Engrs* **5**, 251-259.

Lindberg, N.-O. (ed.) (1988) *Acta Pharm. Suec.* **25**, 185-280 (Special Issue).

Lindberg, N.-O., Jönsson, C. and Holmquist, B. (1985). *Drug Dev. Ind. Pharm.* **11**, 917-930.

McTaggart, C. M., Ganley, J. A., Sickmueller, A. and Walker, S. E. (1984). *Int. J. Pharm.* **19**, 139-148.

Newitt, D. M. and Conway Jones, J. M. (1958). *Trans. Inst. Chem. Eng.* **36**, 422-442.

Ormós, Z., Csukás, B. and Pataki, K. (1975). *Hung. J. Ind. Chem.* **3**, 631-646.

Ouychiyama, N. and Tanaka, T. (1982). *Ind. Eng. Chem. Process. Des. Dev.* **21**, 35-37.

Paris, L. and Stamm, A. (1985). *Drug Dev. Ind. Pharm.* **11**, 333-360.

Parker, M. D., York, P. and Rowe, R. C. (1990). *Int. J. Pharm.* **64**, 207-216.

Parker, M. D., York, P. and Rowe, R. C. (1991). *Int. J. Pharm.* **72**, 243-249.

Pataki, K., Horváth, E. and Ormós, Z. (1983). *Proc. 4th Conf. Appl. Chem. Unit Oper. Processes* **3**, 258-267.

Pietsch, W. B. (1969). *J. Eng. Ind.* **91**, Ser. B (No. 2), 435-449.

Pietsch, W. B. (1984). In *Handbook on Powder Science and Technology* (Fayed, M. E. and Otten, L., eds) pp. 231-252. Van Noostrand Reinholdt Co., New York.

Richardson, B. (1982). *Proceedings of the Interphex conference 1982*, Brighton, UK.

Ritala, M., Jungersen, O., Holm, P., Schaefer, T. and Kristensen, H. G. (1986). *Drug Dev. Ind. Pharm.* **12**, 1685-1700.

Ritala, M., Holm, P., Schaefer, T. and Kristensen, H. G. (1988). *Drug Dev. Ind. Pharm.* **14**, 1041-1060.

Rowe, R. C. (1988). *Acta Pharm. Technol.* **13**, 144-146.

Rowe, R. C. (1989a). *Int. J. Pharm.* **52**, 140-154.

Rowe, R. C. (1989b). *Int. J. Pharm.* **53**, 75-78.

Rowe, R. C. (1989c). *Int. J. Pharm.* **56**, 117-124.

Rowe, R. C. (1990). *Int. J. Pharm.* **58**, 209-213.

Rubinstein, M. H. (1976). *J. Pharm. Pharmacol.* **28** Suppl., 67P-71P.

Rumpf, H. (1962). In *Agglomeration* (Knepper, W. A., ed.), pp. 379-414. Interscience Publishers, New York.

Rumpf, H. and Turba, E. (1964). *Ber. Dtsch. Keram. Ges.* **41** (2), 78-84.

Sastry, K. V. S. and Fuerstenau, D. W. (1977). In *Agglomeration 77* (Sastry,

K. V. S., ed.), pp. 381–402. AIME, New York.

Schaefer, T. (1988). *Acta Pharm. Suec.* **25**, 205–228.

Schaefer, T. and Woerts, O. (1978a). *Arch. Pharm. Chemi, Sci. Ed.* **6**, 1–13.

Schaefer, T. and Woerts, O. (1978b). *Arch. Pharm. Chemi, Sci. Ed.* **6**, 14–25.

Schaefer, T., Holm, P. and Kristensen, H. G. (1986a). *Arch. Pharm. Chemi, Sci. Ed.* **14**, 1–16.

Schaefer, T., Bak, H. H., Jaegerskou, A., Kristensen, A., Svensson, J. R., Holm, P. and Kristensen, H. G. (1986b). *Pharm. Ind.* **48**, 1083–1089.

Schaefer, T., Bak, H. H., Jaegerskou, A., Kristensen, A., Svensson, J. R., Holm, P. and Kristensen, H. G. (1987). *Pharm. Ind.* **49**, 297–304.

Schaefer, T., Holm, P. and Kristensen, H. G. (1990a). *Pharm. Ind.* **52**, 1147–1153.

Schaefer, T., Holm, P. and Kristensen H. G. (1990b). *Drug Dev. Ind. Pharm.* **16**, 1249–1277.

Schaefer, T., Holm, P. and Kristensen, H. G. (1992a). *Acta Pharm. Nord.* **4**, 133–140.

Schaefer, T., Holm, P. and Kristensen, H. G. (1992b). *Acta Pharm. Nord.* **4**, 141–148.

Schaefer, T., Holm, P. and Kristensen, H. G. (1992c). *Acta Pharm. Nord.* **4**, 245–252.

Schaefer, T., Taagegaard, B., Thomsen, L. J. and Kristensen, H. G. (1993a). *Eur. J. Pharm. Sci.* **1**, 125–131.

Schaefer, T., Taagegaard, B., Thomsen, L. J. and Kristensen, H. G. (1993b). *Eur. J. Pharm. Sci.* **1**, 133–141.

Schubert H. (1975). *Powder Technol.* **11**, 107–119.

Sherington, P. J. and Oliver, R. (1981). *Granulation.* Heyden & Son, London.

Shirakuna, D., Yamada, M., Hashimoto, M., Ishimaru, S. and Nagai, T. (1992). *Drug Dev. Ind. Pharm.* **18**, 1099–1110.

Takenaka, H., Kawashima, Y. and Hishida, J. (1981). *Chem. Pharm. Bull.* **29**, 2653–2660.

Tapper, G.-I. and Lindberg, N.-O. (1986). *Acta Pharm. Suec.* **23**, 47–56.

Thomsen, L. J. (1992). *Matrix pellets. Ph.D. thesis*, The Royal Danish School of Pharmacy, Copenhagen.

Thomsen, L. J., Schaefer, T., Møller-Sonnergaard, J. and Kristensen, H. G. (1993a). *Drug Dev. Ind. Pharm.* **19**, 1867–1887.

Thomsen, L. J., Schaefer, T. and Kristensen, H. G. (1993b). *Drug Dev. Ind. Pharm.* **20**, 1179–1197.

Thomsen, L. J. (1994). *Pharm. Tech. Eur.* **6** (9), 19–24.

Tsubaki, J. and Jimbo, G. (1984). *Powder Technol.* **37**, 219–227.

Ullah, I., Corrao, R. G., Wiley, G. J. and Lipper, R. A. (1987). *Pharm. Tech.* **11**, 48–54.

Vojnovic, D., Seleneti, P., Rubessa, F., Moneghini, M. and Zanchetta, A. (1992). *Drug Dev. Ind. Pharm.* **18**, 961–972.

Wehrle, P., Nobelis, D., Cuine, A. and Stamm, A. (1992). *STP Pharma Practique* **2**, 38–44.

Werani, J. (1988). *Acta Pharm. Suec.* **25**, 247–266.

Zhang, G., Schwartz, J. B. and Schnaare, R. L. (1990). *Drug Dev. Ind. Pharm.* **16**, 1171–1184.

INDEX